Exemplary Science
for Resolving
Societal Challenges

Exemplary Science for Resolving Societal Challenges

Edited by Robert E. Yager

National Science Teachers Association

Arlington, Virginia

National Science Teachers Association

Claire Reinburg, Director
Jennifer Horak, Managing Editor
J. Andrew Cooke, Senior Editor
Judy Cusick, Senior Editor
Wendy Rubin, Associate Editor
Amy America, Book Acquisitions Coordinator

ART AND DESIGN, Will Thomas, Jr., Director

PRINTING AND PRODUCTION
Catherine Lorrain, Director
Nguyet Tran, Assistant Production Manager

NATIONAL SCIENCE TEACHERS ASSOCIATION
Francis Q. Eberle, PhD, Executive Director
David Beacom, Publisher

LIBRARY OF CONGRESS CATALOGING-IN-PUBLICATION DATA
Exemplary science for resolving societal challenges / edited by Robert Yager.
   p. cm.
 Includes index.
 ISBN 978-1-936137-12-1
 1. Science—Study and teaching—Standards—United States. 2. Science—Social aspects—United States. I.
Yager, Robert Eugene, 1930-
 Q183.3.A1E94 2010
 303.48'3071—dc22
                              2010022947
eISBN 978-1-936137-60-2

NSTA is committed to publishing quality materials that promote the best in inquiry-based science education. However, conditions of actual use may vary and the safety procedures and practices described in this book are intended to serve only as a guide. Additional precautionary measures may be required. NSTA and the author(s) do not warrant or represent that the procedure and practices in this book meet any safety code or standard or federal, state, or local regulations. NSTA and the author(s) disclaim any liability for personal injury or damage to property arising out of or relating to the use of this book including any recommendations, instructions, or materials contained therein.

PERMISSIONS

Contents

Using Two of the Goals for School Science and One Category of Science Content as Advocated in the National Science Education Standards (NSES)

Robert E. Yager
University of Iowa

T he final version of the National Science Education Standards (NSES) was published in 1996 after four years of intensive debate, several trial editions, and the expenditure of seven million dollars of public funds. The focus for this monograph arises from two aspects of the NSES that too often are ignored as reforms are urged. One of these concerns two of the four goals that should frame reform efforts designed in science for preK–12 schools. The Standards also define eight categories of science content standards, enlarging the "playing field" beyond the traditional science "disciplines." The content category that is the major focus of this monograph is that students should have experiences with personal and social issues where the constructs of science can be used.

None of the goals specify content typically included in textbooks and the curriculum. Similarly, they have not affected state standards used to indicate what most achievement tests measure. The Standard's four goals for school science are to prepare students to

1. experience the richness and excitement of knowing about and understanding the natural world;

2. use appropriate scientific processes and principles in making personal decisions;

3. engage intelligently in public discourse and debate about matters of scientific and technological concerns; and

4. increase their economic productivity through the use of the knowledge, understandings, and skills of the scientifically literate person in their careers (NRC 1996, p. 13).

These are the goals we are asked to use as we change teaching, plan the continuing education of teachers, and consider ways to assess student learning. This monograph includes examples of actions designed to meet Goals 2 and 3 and the content focus on use of science skills and constructs to deal with personal and societal challenges.

Much attention is needed to prepare all citizens (including students of all school ages) to become involved with the problems and issues that affect human existence in homes, schools, and local government. This monograph was conceived to illustrate the centrality and importance of both Goals 2 and 3 and to make every teacher responsible for helping students improve their own lives as well as society in general. The goals and the content focus on realizing personal and societal problems are infrequently considered as worthy of attention or used to illustrate the use and understanding of science for all learners. These all are critical for a democracy to work! They also exemplify what is needed before reforms for school science can be undertaken.

Too often education goals go unnoticed or are merely conceived as broad statements used by administrators and state leaders to "glorify" teaching and learning. Few educators view the specific goals as something to consider before framing a curriculum, choosing instructional materials, and selecting instruments to assess student performance. Many teachers who are aware of the NSES have never read nor internalized the discussion of the goals and how they could and should be approached in their teaching. The work of Wiggins and McTighe (1998) is important in this regard. They urge all teachers to spend time in developing goals with their students and then immediately discussing and agreeing to the kinds of evidence that they could collect to indicate that the goals have been met. This is why assessment comes after teaching but *before* considering content in the NSES. Textbooks seldom, if ever, refer to every goal proposed in the NSES or to all eight categories of the science content standards.

The NSES do provide help with defining science content. Eight science content categories are recommended for use in meeting the goals in preK–12 schools. Once more there is little discussion and few examples of all eight in school curricula, state standards, or textbooks. Too many merely outline concepts anew from the basic disciplines of science, namely physics, chemistry, biology, and Earth and space science. Many chose to use NSES recommendations but ignore the unique and uncommon ones. The eight categories of content standards for science are

1. unifying concepts and processes in science,

2. science as inquiry,

3. physical science,

4. life science,

5. Earth/space science,

6. science and technology;

7. science in personal and social perspectives, and

8. history and nature of science.

The NSES recommend the consideration of all eight of these science content standards.

Inquiry was initially introduced in the early 1960s as a major new focus for school science. It was a new idea and even somewhat controversial. But today even textbooks claim a focus on inquiry. Another change introduced in the reforms of the 1960s was a focus on process skills. The American Association for the Advancement of Science (AAAS) developed a major K–8 program making process skills central to school science programs. Many continue to list the 14 skills scientists used to illustrate the meaning for inquiry. But when processes frame the curriculum completely, little reform is actually noted. That is why unification of concepts and skills is recommended first and as a "form" of content. Didactic teaching can succeed even when concepts and processes are defined and used to indicate specifically what students need to recite, though little real understanding and use usually occurs.

Technology is a word that remains confusing for many. Too many view it as computer technology, rather than a whole field dealing with the human-made world, encompassing engineering, medicine, and invention. In the 1960s Zacharias (the architect of the first of the alphabet courses, namely PSSC), advocated getting rid of all technology in textbooks, because "it was not science!" Now many see technology as being more interesting and vital for students than basic science. Further, in the real world of science there is major dependence on technology; it enables much science to be undertaken. This is very different from defining technology as the "applications of science."

The seventh category in the content standards—a focus on science from personal and social perspectives—is the theme of this book. Of special interest is the fact that such "content" is included for all three grade level groupings used in the NSES. It is also the means for illustrating ways that Goals 2 and 3 can be approached, as the content of science must include science from both personal and social perspectives.

This monograph includes 15 chapters written by diverse groups of writers, educators, and scientists who report on situations where Goals 2 and 3 are seriously considered and where science content is approached from personal and societal perspectives. The monograph illustrates how personal and social contexts have been approached in ways not found in mainline curricula or in the most-used science textbooks. Hopefully the 15 examples that follow will provide a new look at the Standards and encourage a broader view of science content anchored in our world. This can be done with less concern for merely covering the typical discipline topics. The authors report on exciting new strategies developed for meeting the goals while also considering the specific content recommendations central to reform. The chapters are diverse but all provide examples of real change and real reform.

References

American Association for the Advancement of Science (AAAS). 1990. *Science for all Americans*. New York: Oxford University Press.

National Research Council (NRC). 1996. *National science education standards*. Washington, DC: National Academies Press.

Wiggins, G. P., and J. McTighe. 1998. *Understanding by design*. Alexandria, VA: Association for Supervision and Curriculum Development.

Zacharias, J. 1956. *Physical sciences study committee (PSSC)*. Cambridge, MA: Massachusetts Institute of Technology Press.

Acknowledgments

Members of the National Advisory Board for the Exemplary Science Series

Lloyd H. Barrow
Missouri University Science
Education Center Member
Professor
Science Education
University of Missouri
Columbia, MO 65211

Bonnie Brunkhorst
Past President of NSTA
Professor of Geological Science and
Science Education
California State University –
San Bernardino
San Bernardino, CA 92506

Lynn A. Bryan
Professor of Science Education
Department of Curriculum and
  Instruction
Purdue University
West Lafayette, IN 47907

Charlene M. Czerniak
Professor of Science Education
Department of Curriculum and
  Instruction
University of Toledo
Toledo, OH 43606

Linda Froschauer
NSTA President 2006–2007
Editor, *Science & Children*
NSTA
Arlington, VA 22201

Stephen Henderson
Vice President for Education Programs
Kentucky Science and Technology
Corporation
Lexington, KY 40506

Bobby Jeanpierre
Associate Professor
College of Education
University of Central Florida
Orlando, FL 32816

Janice Koch
Professor Emerita
Science Education
Department of Curriculum and Teaching
Hofstra University
Long Island, NY 11549
Mailing address:
7843 Maple Lawn Blvd
Fulton, MD 20759

LeRoy R. Lee
Executive Director
Wisconsin Science Network
4420 Gray Road
De Forest, WI 52532-2506

Shelley A. Lee
Science Education Consultant
WI Dept. of Public Instruction
PO Box 7842
Madison, WI 53707-7841

acknowledgments

Edward P. Ortleb
Science Consultant/Author
5663 Pernod Avenue
St. Louis, MO 63139

Carolyn F. Randolph
Science Education Consultant
14 Crescent Lake Court
Blythewood, South Carolina 29016

Barbara Woodworth Saigo
President
Saiwood Publications
23051 County Road 75
St. Cloud, MN 56301

Patricia Simmons
Professor and Department Head
Math Science & Technology Education
North Carolina State University
Raleigh, NC 27695

Gerald Skoog
Texas Tech University
College of Education
15th and Boston
Lubbock, TX 79409-1071

Vanessa Westbrook
Director, District XIII
Senior Science Specialist
Charles A. Dana Center
University of Texas at Austin
Austin, TX 78722

Mary Ann Mullinnix
Assistant Editor
University of Iowa
Iowa City, Iowa 52242

About the Editor

Robert E. Yager—an active contributor to the development of the National Science Education Standards—has devoted his life to teaching, writing, and advocating on behalf of science education worldwide. Having started his career as a high school science teacher, he has been a professor of science education at the University of Iowa since 1956. He has also served as president of seven national organizations, including NSTA, and has been involved in teacher education in Japan, Korea, Taiwan, Indonesia, Turkey, Egypt, and several European countries. Among his many publications are several NSTA books, including *Focus on Excellence* and two issues of *What Research Says to the Science Teacher*. He has authored over 600 research and policy publications as well as having served as editor for seven volumes of NSTA's Exemplary Science Programs (ESP). Yager earned a bachelor's degree in biology from the University of Northern Iowa and master's and doctoral degrees in plant physiology from the University of Iowa.

Developing Expertise in Project-Based Science:

A Longitudinal Study of Teacher Development and Student Perceptions

Gail Dickinson, Emily J. Summers, and Julie K. Jackson
Texas State University–San Marcos

Defining Project-Based Science: A Contextual History

While Project-Based Science (PBS) has deep theoretical traditions starting at the turn of the last century, it clearly relates to Goals 2000 Objective 4 and the subsequent formation of the National Standards. PBS began as an extension of the American progressive education movement of the early 1900s when Kirkpatrick (1918) asserted that education should center on children engaging in self-directed, purposeful projects. Dewey (1938) concurred that real-world connections were important in the learning process; however, he contended that projects needed to take the form of joint ventures between teachers and children. Vygotsky's (1962) social development theory argued that language and cognition were inextricably linked. This, along with new findings in the cognitive sciences formed the basis of a transition from individual, competitive learning environments to cooperative learning environments (Slavin 1985).

PBS provides a classroom interface that integrates Dewey's real-world problem solving opportunities with Slavin's cooperative learning strategies. We define PBS as "an instructional method using complex, authentic questions to engage students in long-term, in-depth collaborative learning, resulting in a carefully designed product or artifact" (Dickinson and Jackson 2008). Authentic PBS projects have the following characteristics: (a) a driving question or problem scenario that fosters sustained student engagement over weeks or months, (b) content that is central to curricular standards and meaningful to students, (c) collaboration between students and/or students and adults, (d) ongoing assessment of student work, and (e) student-designed investigations (Krajcik, Czerniak, and Berger 2002; Thomas and Mergandoller 2000).

With PBS emphasis on student work products, public presentations, and projects aligned to address community issues, it strongly aligns to and supports Goals 2000 Objective 4, *students will engage intelligently in public discourse and debate about matters of scientific and technological concern* (Krajcik, Czerniak, and Berger 2002; Steinberg 1997). Moreover, research indicates that PBS increases typical student science achievement (Marx et al. 2004), increases scientific inquiry skills (Baumgartner and Zabin 2008), provides students with a more holistic view of the discipline (Boaler 2002), and engages students who struggle in most academic settings (Wurdinger et al. 2007).

Despite offering promise, PBS presents unique challenges for teachers and students:

- PBS requires deep and flexible teacher content knowledge.
- PBS requires more effort for both the teacher and students.
- PBS is time consuming.
- Classroom management is more difficult in PBS than transmission approaches to instruction.
- Teachers must provide adequate scaffolding for students to succeed without stifling student investigations.
- Students feel more comfortable in traditional classes than PBS (Ladewski, Krajcik, and Harvey 1994; Beck, Czerniak, and Lumpe 2000; Frank and Barzilai 2004).

Toolin (2004) finds that teachers with strong content and pedagogy backgrounds are more likely to implement projects in their classes than those who lack formal training in education. She also asserts that first-year teachers with support structures (such as team teaching, one-on-one professional development, and professional development workshops) become capable of implementing successful PBS units. We contend that preservice training in PBS facilitates early faithful implementation of PBS, which in turn provides students with the opportunity to engage in the public discourse and debate advocated in Goals 2000 Objective 4.

Introduction

This multifaceted study took place over an eight-year time span and covers two educational settings. We report on the participants as both preservice and inservice teachers. Additionally, we report on the learning and perceptions of the secondary students with whom the participants currently teach.

The Preservice University Program Setting

Participants in this study completed an undergraduate teacher preparation program that culminated with a project-based instructional course. This capstone course included four key elements: reading about PBS, making field observations of established PBS classrooms, sharing teaching experiences, and designing a PBS unit (See Figure 1.1).

Figure 1.1. The Relationship Between Components of the PBS Course

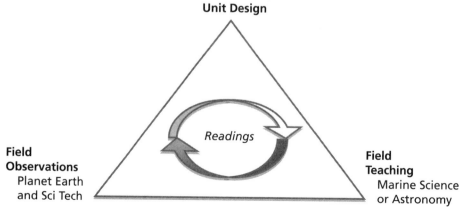

Unit Design

Readings

Field Observations
Planet Earth
and Sci Tech

Field Teaching
Marine Science
or Astronomy

Readings on and Field Observations of PBS

Preservice science teachers read Polman's (2000) account of a seasoned high school teacher who struggled implementing PBS. Students also observed four hours of PBS instruction at a local high school. The PBS classes included *Planet Earth*, an interdisciplinary course on the origins and evolution of the Earth and human impacts on the Earth, and *Science and Technology*, a physical science course modeled after the Massachusetts Institute of Technology Mousetrap Challenge. After each classroom observation, the preservice teachers posted neutral observation reports in online community forums. Reading reflections and posted observation comments formed the basis for class discussions about the benefits and limitations of PBS, management of PBS environments, and what constitutes PBS. Additionally, these community forums provided valuable data collection opportunities.

Teaching Experiences

All preservice participants chose either a marine science- or astronomy-focused teaching experience. The marine science teaching experience option involved two weekend field trips to a marine institute located on the coast about 250 miles from the university. The first field trip was designed to orient preservice teachers to the facilities, coastal environments, and possible teaching topics. At the end of the first field trip, preservice teachers brainstormed ideas and identified a driving question for the next field trip. They spent about a month organizing lessons around that driving question. The second field trip was a joint venture with several high schools. During this second field trip, preservice teachers taught inquiry lessons to secondary students with the goal of collaboratively addressing the driving question for the trip. For example, the driving question for one trip was, "Is the marine environment an opportunity for living organisms to exploit or an obstacle to be overcome?" Lessons taught on the jetty emphasized the challenges of living on a hard surface with pounding surf and tides whereas lessons taught at the salt marsh emphasized adaptations for living in an anaerobic soft substrate with almost no wave action. Culminating lessons encouraged debate about the driving question and human impacts and responsibilities for protecting these environments.

The astronomy teaching experience option also involved two all-day, in-school field trips at a local high school. The driving question, "How can we use mathematics to design and use a Dobsonian telescope?" was provided by the instructor. Preservice teachers built the bases of Dobsonian telescopes and then taught lessons that included defining a parabola, using conic sections to identify the focal length of the primary mirror (Siegel, Dickinson, and Hooper 2008), discovering the mathematical basis for light reflection on straight and parabolic mirrors, and positioning mirrors and eyepieces within the telescope tube. Through these lessons, the high school students explored properties of light and parabolas while constructing the rest of the telescope. While this field option primarily targeted preservice mathematics teachers in the course, many preservice science teachers opted for this option.

Curriculum Design

To prepare preservice teachers philosophically and pedagogically for the teamwork aspects of PBS teaching, participants worked in teams of two or three to develop a four-to-six week project-based unit that included a driving question, concept map, project calendar, selected lesson plans, a three-to-five minute anchor video, assessments, grant proposal, resource list, modifications for special needs students, and a short paper introducing the project to their peers. Preservice teachers were strongly encouraged to develop projects that fostered public discourse of socioscientific issues.

Scaffolding Curricular Unit Design. Development of the curriculum unit was highly structured and included the expectation that preservice teachers would revise each part until it met the standard for acceptance. At the beginning of the semester, preservice teachers were given a rubric that identified and defined the unit components (see Figure 1.2). Toward the end of the semester, professors provided the preservice teachers with an HTML template for the project. The template was a folder with HTML files for each project component. Each HTML file was set up as a table with a navigation bar on the left, a field at the top for the unit title and authors, and a field on the right for the unit component. Preservice teachers converted document and concept map files into HTML or graphics files and pasted them into the fields on the template files. When they completed filling in their HTML templates, we compiled the units into a class CD and posted them online for future reference at *www.education.txstate.edu/ci/faculty/dickinson/ PBI*. The template provided uniformity among the projects and made the projects accessible to preservice teachers across semesters and among graduates. Additionally, using a template limited the technology skills required. This put the emphasis on the content rather than the technology. Nonetheless, preservice teachers acquired some technology skills in the process.

Developing High-Quality Driving Questions. Several class sessions were devoted to defining PBS, analyzing sample PBS driving questions for quality using Krajcik, Czerniak and Berger's (2002) five criteria for driving questions, and providing diverse examples of PBS including units from previous semesters. After examining the curriculum of the largest school district in the area, each preservice teacher devised a driving question and an explanation of how the driving question was a good one for the targeted knowledge and skill standard, grade level, and discipline. The preservice teachers posted these online for classmates to review. After online peer reviews, the preservice teachers revised their driving questions. Preservice teachers then selected questions from their unit from the list of driving questions that had been generated as a whole. Usually, about one-third of the driving questions were of sufficient quality for the units so preservice teachers typically worked in groups of three to develop their units.

To further assure that the driving questions were sustainable and central to the curriculum, preservice teacher teams developed concept maps of their driving questions. They correlated their maps with state standards and local district curricula to see if the unit was feasible in a school setting (For example, did the unit cover sufficient numbers of the state standards to be worthwhile for the amount of time devoted to the unit? A sufficient number of standards would require a pace that allowed for most or all of the state standards to be met in the context of the course).

Developing an Anchor Video for the Unit. Ideally, developing a unit calendar would come next; however, because video cameras and editing equipment were typically in high demand at the end of the semester, preservice teachers developed a short anchor video before developing the rest of the unit. We used iMovie to edit and compress the videos because it is very intuitive and has an excellent tutorial. Preservice teachers typically developed one of three types of video: narrated slide show, skit, or video montage. The videos also varied in how much

Figure 1.2. Sample Rubric for Project-Based Unit

	Points	Accepted (on time and meets standards)	Needs Revision (on-time but needs some revision (-15%)	Needs Conference/ Late (-30%)
Concept Map (2/9) Constructed in Inspiration and saved in HTML format in template file named map.htm	5			
Anchor Video Storyboard (2/11)	5			
Anchor Video (2/21) Quicktime format. Compress as CDROM Med and saved as movie.mov. Compress a second copy as Web small and save as smmovie.mov.	10			
Lesson Plans (2/25) use template from Step II. Save as HTML in files titled lesson1.htm, lesson2.htm, etc.	10			
Resources (2/25) HTML format in template file named resource.htm Web resources (3 annotated URLS per person) Print materials list Supplies (labware, hardware, software, etc.,)	10			
Project Calendar (3/4) HTML format in template file named calendar.htm	5			
Grant (4/15) Use TAPESTRY form. Hard copy and HTML format in template file named grant.htm	10			
Letter to Parents (4/22) HTML format in template file named parents.htm	5			
Assessments (4/22) HTML format in template file named assess.htm	10			
Modifications for Special Needs (4/22)	10			
Introductory Paper (4/29) HTML format in file titled intro.htm Target Audience Project description (1 paragraph) Driving question Overall goals of the project Project objectives Rationale—include theoretical reasons for doing the project Background—1–2 pages of background info (content specific) Standards addressed TEKS National Standards (NCTM or NSTA) National Technology Standards Description of formative and summative assessments including description of final product	10			
Final Presentation (final exam period)	10			
Total Points	100			
All group members participate equally. If not, grades will be weighted appropriately.				

information they provided students. Some videos led students to generate their own questions while others were more prescriptive, providing students with a single driving question they would answer.

Unit Calendar, Lesson Plan, and Assessment Development. The next step was developing a unit calendar. The calendar provided another check for sustainability and curricular centrality. If preservice teachers were unable to plan a four-week unit including engaging classroom activities that supported deep understanding of key concepts, they needed to revise their driving questions. If the driving questions covered too much of the curriculum for the scope of the unit, preservice teachers either scaled back their driving questions or selected a 4–6 week part of the unit as their focus. Each preservice teacher selected two lessons from his or her unit calendar to flesh out in lesson plans. Preservice teachers revised their calendars to include diverse, ongoing assessments derived from *Classroom Assessment Techniques* (Angelo and Cross 1993).

The Inservice Public School Program Setting

The southwestern U.S. school district used as the field site for inservice teacher observations was, at the time of study, ethnically diverse with the following student population: 69.5% economically disadvantaged, 56% Hispanic, 28% African American, and 14% White. The district had two high schools: one centered on a PBS curriculum and the other utilizing a traditional curriculum. We chose preservice teachers who took teaching positions in a campus where the greatest percentage of time was spent teaching PBS because this setting uniquely allowed us to examine the development of teacher expertise and student outcomes.

PBS in Biology. The inservice biology teachers developed a PBS unit around the driving question, "Are genetically modified foods good or bad?" They constructed a rubric that guided students through aspects of genetics necessary to understand the debate concerning genetically modified organisms. Because students were unaware which side they would represent until a few days before the actual debate, all teams researched both the pros and cons of this issue. Pitting student teams from one class against another added a motivating competitive element.

The inservice biology teachers also developed a PBS unit on speciation that included the driving question, "How would a species adapt to a specific speciation event over thousands of generations?" To begin this project, high school students select a current species and then drew a speciation event out of a hat. Students imagine how their species might adapt in an evolutionary sense. They created a field guide for the new species; they assign it a name, talk about its adaptations, give a phylogeny of where it fits with the previous species, create a dichotomous key to identify it, and sketch a picture of it.

PBS in Astronomy. The astronomy inservice teachers developed a project where students role-play groups of scientists gathering for a conference on misconceptions about the seasons. The groups represent cities at different latitudes so students must share data across groups to look for patterns that will support scientific explanations about the cause of the seasons. The teacher recruits scientific experts from the community to participate in the conference as audience members for student presentations. The experts quiz students and judge student presentations. This motivates students to prepare thoughtful explanations and exceed project requirements.

Methodology

Participants

This study used a sample of 121 inservice teachers who completed exit surveys at the end of their undergraduate program, 62 inservice teachers who completed longitudinal surveys marking the preservice program's 10-year anniversary, longitudinal preservice teacher observations of two PBS classrooms, interviews and surveys with a random stratified sample of 12 students enrolled in PBS science classes, and in-depth interviews and three years of ethnographic observations of three inservice PBS teachers from their time in an intensive preservice PBS teacher training through their second year of teaching in a PBS-focused high school. The randomly selected student sample was stratified and weighted to emphasize the district's at-risk populations (75% Hispanic, 25% African American, and 17% White).

PBS in Action: A Purposeful In-Depth Look at Three Exemplar PBS Teachers

The teachers chosen as exemplars for the longitudinal ethnographic portion of this study actively sought an educational environment that supported their personal philosophies of teaching. They were recruited from the capstone PBS course by the school principal prior to student teaching. All three teachers included in this qualitative profile of PBS in action completed the program as post-baccalaureates. Stacy (all names pseudonyms) was a non-degree-seeking post-baccalaureate with a BS in environmental science. Christine was earning her masters degree in biology while she completed the program and Jackie was a PhD candidate in Physics. Stacy teaches biology and chemistry, Christine teaches biology and Jackie teaches astronomy and physical science.

Data Sources

Data sources included preservice teacher program exit surveys, preservice teacher program graduate surveys, classroom observations, teacher interviews, student interviews and student surveys. Exit interviews were routinely conducted as part of the preservice teacher program evaluation. These interviews provided a snapshot of preservice teacher perceptions at graduation. To ascertain practicing teacher perceptions, we surveyed program graduates by mail. We observed case study teachers over a period of two years and interviewed them at the end of their second year of teaching. We also surveyed and interviewed high school sophomores at the end of the school's second year to determine their perceptions of the project-based environment.

Analyses

Interview data were transcribed and coded. We analyzed interview and observation data identifying five themes that describe students' perceptions of PBS: (a) success, (b) differences in PBS from other previous science instruction, (c) gender differences, (d) public schooling contexts and barriers to embracing change, and (e) real-life connections. Surveys were descriptively analyzed because of the small sample. Observations were recorded as thick descriptions and coded. We used SPSS (version 15.0) for statistical analyses.

Findings

Preservice Teaching Programmatic Findings

According to the inservice teachers whom we interviewed and/or surveyed, the most significant aspects of the preservice PBS training program were development of the PBS unit, production of an anchor video, and use of the *Classroom Assessment Techniques* text. Inservice teachers kept copies of *Classroom Assessment Techniques* in their classrooms for ready reference and mentioned using it often.

Preservice Program Exit Survey Findings. We utilized the exit survey data for a random selection of the eight years of available science preservice teacher data to provide a contextual backdrop to understand the teacher participants' attitudinal changes as they transitioned from preservice to inservice PBS teaching. The capstone PBS course included preservice teachers in mathematics, computer science, and science, but we limited the reported sample in this chapter to preservice science teachers to align to the inservice teacher case studies. As a whole, the preservice science teachers had significantly higher levels of PBS teaching confidence than the mathematics or computer science teachers (.038). Upon graduation, teachers' PBS teaching confidence was not statistically different from other areas of teaching confidence such as inquiry teaching (1.512), science teaching (2.53), direct teaching (2.53), or teaching confidence (.55). We examined multiple areas of teaching confidence for the entire sample of preservice teachers and exclusively for the preservice science teachers in the sample (see tables 1.1 and 1.2).

Table 1.1. Descriptive Statistics of Preservice Teachers' Teaching Confidence at Time of Graduation

	N	Range	Minimum	Maximum	Mean	Std. Deviation
Science Teaching Confidence	23	3	2	5	4.35	.935
PBS Teaching Confidence	23	4	1	5	3.57	.945
Direct Teaching Confidence	23	3	2	5	4.22	.850
Inquiry Efficacy	23	3	2	5	3.96	.878
Collaborative Teaching Confidence	23	3	2	5	4.13	.968
Small Group Teaching Confidence	23	3	2	5	3.96	.976
Student Teaching Confidence	23	1	4	5	4.83	.388

Table 1.2. Descriptive Statistics of Preservice Science Teachers' Teaching Confidence at Time of Graduation

	N	Range	Minimum	Maximum	Mean	Std. Deviation
Science Teaching Confidence	9	2	3	5	4.44	.726
PBS Teaching Confidence	9	3	2	5	3.56	.882
Inquiry Efficacy	9	2	3	5	3.78	.667
Direct Teaching Confidence	9	1	4	5	4.33	.500
Collaborative Teaching Confidence	9	3	2	5	3.67	.866
Small Group Teaching Confidence	9	3	2	5	3.89	.928
Student Teaching Confidence	9	1	4	5	4.78	.441

Inservice Teaching

Longitudinal Survey Findings

We mailed program evaluation surveys to a sample (n = 30) representing graduates from 10 years of the program. Two-thirds of the respondents (n = 20) completed the program as undergraduates. The remaining respondents (n = 10) completed the program as post-baccalaureates. The respondents had been teaching an average of 2.45 years (sd = 1.35).

Table 1.3. Demographics of Surveyed Inservice Teachers

Type of School	N
Rural High School	1
Private High School	1
Suburban High School	5
Urban High School	13
Suburban Middle School	2
Urban Middle School	8
Status During Certification Program	
Undergraduate	20
Graduate	10
Number of Years Teaching	
1	9
2	5
3	8
4	4
5	4

Upon entering full-time inservice teaching, 22 of the 30 graduates surveyed by mail indicated they used PBS in their teaching. Even so, many indicated problems with implementing PBS. Most identified the lack of instructional and preparation time for extended projects or fear about classroom management as their chief concern. A third-year teacher related the following:

> Project-based instruction would be wonderful, but you are restricted by time and the curriculum you must get through. Also, it is a bit scary because students could come up with questions that are unanswerable or beyond their abilities. I suppose, you could structure and guide the inquiry, as with the CD-ROM [we] made. I will try it this year.

A first-year high school teacher indicated, "Also, many of my students just waste a large amount of time when I try extended projects." Another issue was pressure to conform to perceived cultural norms. One fourth-year middle school science teacher indicated, "Teachers in the same grade/subject teach in a 'traditional' way and [the] administrators expect grade level/subjects to do the same things so there will not be parent phone calls/concerns." Several of the graduates also indicated they would never have known about PBS without having taken the course. One first year science teacher indicated "I try to make PBS as much part of my teaching as possible, and I would not have known how to do that without the PBS course."

Teacher Case Study Findings

School Culture. All three teachers indicated that collaboration with other teachers and support from administrators is key to PBS implementation. Stacy said, "In the program I was in, we had to do things where we worked together and I think the staff, the other science teachers here, we collaborate all the time." Collaboration among teachers helps them refine their projects. Christine said, "I think having people to talk to and having somebody to bounce ideas off of and stuff like that is really, really crucial." The study school uses a critical friends process to review projects before they are implemented. Teachers present their project ideas to their peers who first indicate what they like about the idea. Their peers then identify potential gaps in the project stating what they "wonder" about. For example, teachers may wonder if the project covers sufficient objectives to be worthwhile. Finally, they brainstorm ideas for the project. Christine highlighted the importance of critical peers.

> And that is tremendously helpful. Just because, when you're in by yourself, you don't really have a lot of perspective on your stuff. And, I think getting used to the idea of my stuff is not perfect. It is not my baby. Cause the projects really do start to feel like your baby. And you're like, "Don't touch the baby. I worked really hard on the baby. Stop saying bad things about the baby." You kind of get in a frame of mind of like projects are dynamic. They always change. It's not the baby. The baby can change and that's all right.

Jackie echoed Christine's sentiments.

> I could not get through a day if I hadn't been a more, really open to collaboration with other teachers and [the preservice program] forced you to do that all the time. And that's another huge thing because prior to that I'm an individual learner like I did everything by myself... So it was good to have that practice and to be ready and open-minded to do that because I think if I had been more in that isolated mindset... I probably would've bombed this experience because you can't do it by yourself.

She added that collaboration during her preservice training changed her mindset. "We're always creating our own stuff. And, I never developed the mindset that I'll produce it all myself 'cause we never did."

Christine felt that her discussions with other preservice teachers helped prepare her for the resistance many of her colleagues at other schools face.

> Discussing PBS with my classmates was valuable, because I did not realize how strong my opinion was until I heard people that disagreed with me. And I didn't realize how unusual my opinion was. I assumed everybody would be like, "Awesome! Heck, yeah!"

She added that supportive school culture is important:

Here I'm really lucky because everybody who works here is even if maybe they don't love it they're on board with it…. But even my other friends that teach, every now and then will be like, "I just don't know about that." …And you have to be prepared to deal with that. And I think, if I were to ever go work somewhere besides a [PBS] school, I would have to face that.

Jackie also identified cultural resistance as a constraint; although, she chose to focus on student resistance. "Cause everyone's forced to do it, I don't have to play that game of 'why are we doing this here as opposed to the other classes?' Like if I were to try to be PBS at a school where not everyone was doing it, I'd get a lot of resistance."

All three exemplar PBS teachers indicated that repeated exposure to PBS prepares students to succeed in project-based environments. Jackie states,

The kids are trained now. Collaboration is difficult. You need a lot of practice with it. They're a lot better at managing projects than my freshmen were last year. A lot of issues that just don't come up any more that came up repeatedly [at first].

Stacy also indicated that the students were much better at using project rubrics to guide their progress during the units.

They have a rubric. And they're a lot better this year at using that kind of like a grocery list is OK; did we do this? Did we do that? Those who were with us last year know that's what's happening.

Time Management. All three teachers identified the time involved with planning and implementing PBS as a concern. Stacy indicated,

We realized way before spring break oh my gosh to be where we need to be by the end of this week, it is crunch time. So about two weeks ago it was a conscious effort. We shifted from a project to the short like two, three-day problem-based scenarios where they were still getting practice with orally presenting stuff but it wasn't like a two-, three-week project. We just didn't have time.

Jackie expressed concern that school administrators had unrealistic expectations for how quickly teachers could cover required content using PBS. She stated, "And I think they're saying is, 'Oh, you're new teachers, you'll just get faster….Fundamentally, this strategy is slower. There's a limit to how fast we can go if you're going for depth and if you're using this style." Christine asked, "How do you fit this really cool idea you have into three weeks?" Christine identified preparation time as an issue saying;

But I mean especially with projects, you need long periods of uninterrupted time and it ends up being at home. I knew that but I didn't *know* that. I mean I knew it but I wasn't prepared for exactly how much time it was going to be and how long the hours are.

Christina went on to share that after her first year she established "rules" limiting her time outside of class.

I don't work on Saturdays. I used to come home and give myself an hour to an hour and a half where I didn't work. But, now actually I've switched it around. I go home and do everything really fast so that I have free time in the evening. Yeah, you kind of have to make some rules.

Jackie also acknowledged that projects take a long time to develop and they take longer to implement than traditional methods.

All three exemplar teachers indicated that time management is also critical for students engaged in PBS. Stacy advised teachers to have lots of intermediate deadlines to help students pace themselves and monitor their progress. "So all kinds of deadlines like long term, intermediate, and sometimes daily like "show me something!" ... so that I know 'Are you on the right track? Are you even in the ball park? Are you each helping one another?'" Christine and Jackie strongly advocated developing a project calendar and sticking to it. Christine noted,

Moving deadlines is like death to PBL because then they will never finish and you will get off of your pacing and then you will never catch back up because you know you have to establish the expectation of "we are going to do this and this and this and this week", this and this and this next week, and then we're going to be finished by the end of the next one.

Addressing the Standards. The teachers noted that designing projects around the standards produces units that challenge students and prepare them for state assessments. Christine notes that because PBS takes longer than traditional instruction, it is particularly critical to use instructional time efficiently.

And so you really do have to sit down and analyze the standards and say, "What is the actual verb here? What do students have to do? Do they need to identify? Do they need to analyze?"... But to really start with the standards and go up as opposed to like how can I make this fit?

Jackie develops the rubric first and lists the standards on the left column of her rubrics. "And if that rubric is solid, then I can almost be guaranteed that all of the support materials I'll prepare to get them to satisfy the rubric will be aligned as well." The rubrics provide structure and scaffolding for the students. Jackie notes,

Kids still like to see structure. They still like to have some expectations that you can communicate really clearly, really concisely. So at the end of the day, they're not completely gone and floating in space and not knowing what's going on. There's got to be that balance of choice and still some structure I guess.

Many of the inservice teachers kept and reported using the *Classroom Assessment Techniques* text; whereas, they found Polman's narrative intimidating and not applicable to their daily PBS work.

Contextual Findings
Ethnographic Findings. Within one year of opening, the PBS-focused students significantly outperformed their peers at the district's other non–PBS focused high school on state science assessments. This trend continued during the second year (See Table 1.4).

Table 1.4. Percent of Students Meeting State Science Assessment Standards

	2008			2009		
	PBS High School	**Traditional High School**	**State**	**PBS High School**	**Traditional School**	**State**
All Students	80	54	64	86	47	66
African American	63	45	47	81	37	50
Hispanic	70	50	53	81	47	55
White	99	86	81	94	74	82
Economically Disadvantaged	81	47	50	85	45	53

Student Learning Perceptions Findings. Across all ethnicity and gender strata, all but one of the student respondents indicated that they feel "successful in science class" and "learn a lot" when their teachers utilize PBS instruction (See Table 1.5). All but one student indicated he/she liked science and four indicated it was their favorite class. Most students who indicated that they were interested in being "more successful in science" also had reflective goals for future science-related achievement. An overwhelming majority recognized the teaching shift to PBS instruction as being "integral" to their "science success," agreeing that "the hands-on stuff" contributed to science success. Not surprisingly, teacher roles contributed to students' perceptions of success. Teachers supported student successes by "interacting with students," instead of "just saying the words and teaching us," teachers "join you and know where you're coming from and shows [sic] you different ways on how to get it."

Table 1.5. Descriptive Statistics of Student Perceptions About Project-Based Science

	N	Range	Minimum	Maximum	Mean	Standard Deviation
How much do you like science?	12	5	1	5	4.0	1.1
How much do you learn in science?	12	4	2	5	4.0	0.8
How successful are you in science?	9	2	4	5	4.4	0.5

In keeping with the findings of Frank and Barzilai (2004), the sample of students generally indicated that PBS was "harder" than other methods of science instruction, but was "more engaging" with "more correct answers." Compared to non-PBS instruction, students found that, "we just pretty much worked out of textbooks. And now [in PBS] we don't really use our textbooks unless it's for reference." Instead, in PBS, "we usually do projects or demonstrations." Students recalled doing "a lot of worksheets and fill out stuff out of the book" outside of PBS, while PBS classes involved, "doing more labs and more different types of learning styles with our science." In PBS instruction, students report, "I don't think we've seen a worksheet here."

All the students' responses to academic inquiry questions aligned with correct scientific thinking. They overwhelmingly indicated a community approach to solving scientific problems, even shifting from speaking about non-PBS experiences in the first-person singular "I" to making a subtle shift to the first person plural "we" when speaking about the PBS approach. Gender differences were most apparent when discussing the group work aspects of PBS (Carlo, Swadi, and Mpofu 2003). When asked, "Were there any adjustments you had to make as a student to the way science is taught here using PBS?" Males were more direct about the impact of group work in PBS classrooms. Males exclusively perceived group work as being "challenging." Despite this perceived difficulty of "learning how to work in groups," some males responded that they now "preferred groups". Still, most males agreed that they preferred working independently because "you know exactly what you're doing and you don't have to rely on anybody else."

Females tended to enjoy the collaborative nature of the group work (Deeter-Schmelz, Kennedy, and Ramsey, 2002), providing observations such as, "The good thing about it [group work] is that you can depend on other people and you can meet new people and it helps our community like a family cause it's like our house." This response was typical for a segment of female participants who used gendered or domestic images and language to describe the group work consistent with Zubair's (2008) findings about how females metaphorically use language. When asked about working independently, females, at times, expressed surprise at the question such as one response, "Independently? I don't like it 'cause we're able to ask other classmates but we don't like independent. I'm not really an independent person." So while females may have perceived the groups as positive, this structure may also have served to reinforce traditional models of dependency with ascribed gender roles. Male students' responses also revealed some underlying power imbalances attached to PBS group work, such as this male response, "It's kind of complicated because instead of just being in a group of like in a pair, you're with like three other people and the materials; it just gets too many hands in one section of the lab."

In regular instruction classes, males often dominate the discussion and group pairings. This aligned with what male participants offered, highlighting that they liked included "taking leadership." "I tell them we're going to separate it like this. You bring back the materials and we'll study it out and we'll go from there." On the other hand, some females had assertiveness challenges with group work. PBS helped female students have a vehicle to practice balancing their voice within group interaction findings, "You have to know that it's okay to speak out." Providing mixed gender science communities through PBS may help to balance the gender ratio of students who choose STEM majors in higher education. While the group work aspect of PBS created the greatest gender split in the students' responses, most students, regardless of gender, found benefits in both group and independent work saying, "I like doing both. I like working in groups because you get to interact more with other people and you learn from them and you actually get to know more people sometimes. I like working as an individual sometimes because sometimes you're paired up with people that you can't really trust that much because they're not as good as a worker as you might think. But sometimes working by yourself you may get to pick whatever you want to research and you get your research done."

Conclusions

Implications for Teacher Education

Should PBS be taught in an atmosphere of high stakes testing? We think ample evidence supports that PBS should be taught in high stakes testing environments. Although PBS takes more time than traditional methods of instruction, research indicates that PBS students do as well or better on high stakes tests. Schneider, Krajcik, Marx, and Soloway (2002) assert that high school students engaged in PBS outscored the national sample on 44% of NAEP items. Geier et al. (2008) found similar results in their study of urban middle school students engaged in PBS. They found that the effects of participation in PBS units at different grade levels were independent and cumulative. Higher levels of participation resulted in higher levels of achievement. In our study, the PBS high school students outperformed their similarly situated peers on state-mandated achievement tests.

What level of exposure to PBS do preservice teachers need? While this study did not specifically explore varying levels of exposure to PBS, teachers in this study indicated that preservice exposure to PBS was critical for early adoption. Jackie stated,

I would say from a big picture point of view that I was always in a traditional classroom growing up so the fact that I got exposed to something different and got to see it work is a big deal. Cause, had that intermediate experience not happened, I never would have made the leap to PBS even though now it makes perfect sense.

Stacy concurred. "The PBS class showed me that there was a real viable ALTERNATIVE [sic] to traditional classroom teaching. I was hooked on it once I was in that class." The PBS course constituted one-sixth of the 18-hour teacher preparation program—a significant investment of time that required sacrificing deep coverage of other important topics such as special needs students, and reading in the content area.

Why bother if adoption is low? Preservice program faculty often debated if number of graduates implementing justified the time devoted to a PBS program was great. Faculty felt that universities have an obligation to challenge the status quo even if it means low adoption levels. Inservice teachers adopting PBS serve as cases for future teachers. Van Driel, Beijaard, and Verloop (2001) found that cases provide a powerful tool in reform of teaching practices.

Implications for Practice

The inservice teacher participants repeatedly stated that the best way for them to ensure that project content was aligned to the standards was to start with the standards and work backward. Peer review of projects prior to implementation also serves as a check for centrality as well as providing opportunities for interdisciplinary links. Jackie's use of backward curriculum design with a detailed rubric for the final project helped her stay focused during the project so that students met state requirements.

PBS goes hand-in-hand with national science standards. It provides a vehicle to posit the standards in everyday practice and, when PBS is implemented with fidelity, the student achievement results show that the standards work. Teachers who regularly utilized PBS did more than achieve science content success; they created classroom learning environments where a normative culture of collaborative science was the typical, everyday experience. Participants in our study clearly indicated that designing projects around state standards was essential for addressing testing requirements. Yet, our findings went beyond testing successes. Our study showed that through deeply embedded PBS preservice instruction and a continued trajectory of inservice PBS teaching these participants were able to create classroom communities that imitated how science is done in real-world working contexts. The participants used PBS to bridge the gaps between (a) theory and public school actions, (b) real world science and public school learning, and (c) when the standards become *goals* for science education, the standards become *reality* in reflecting actual student achievement. Moreover, our study showed how PBS filled gaps between stated goals and actual student achievements in districts with large pockets of students who were identified with low socioeconomic status, rural, Hispanic, first generation college-bound, and English language learners. A large portion of students surveyed indicated that they will be the first person in their family to graduate with a U.S. high school diploma; yet, like their fellow PBS learners, they held high hopes of studying science beyond high school.

Acknowledgment

The authors would like to acknowledge Michael Marder and Melissa Dodson for research assistance in gathering the end-of-program, undergraduate preservice teacher participant responses.

References

Angelo, T. A., and K. P. Cross. 1993. *Classroom assessment techniques: A handbook for college teachers*. 2nd ed. San Francisco: Jossey Bass.

Baumgartner, E., and C. Zabin. 2008. A case study of project-based instruction in the ninth grade: A semester-long study of intertidal biodiversity. *Environmental Education Research* 14 (2): 97–114.

Beck, J., C. M. Czerniak, and A. T. Lumpe. 2000. An exploratory study of teachers' beliefs regarding the implementation of constructivism in their classrooms. *Journal of Science Teacher Education* 11 (4): 323–343.

Boaler, J. 2002. *Experiencing school mathematics: Traditional and reform approaches to teaching and their impact on student learning*. Mahwah, NJ: Lawrence Erlbaum Associates.

Carlo, M. D., H. Swadi, and D. Mpofu. 2003. Medical student perceptions of factors affecting productivity of problem-based learning tutorial: Does culture influence the outcome? *Teaching and Learning in Medicine* 15 (1): 59–64.

Deeter-Schmelz, D. R., K. N. Kennedy, and R. P. Ramsey. 2002. Enriching our understanding of student effectiveness. *Journal of Marketing Education* 24 (2): 114–124.

Dewey, J. F. 1938. *Experience and education*. Indianapolis: Kappa Delta Pi.

Dickinson, G., and J. K. Jackson. 2008. Planning for success: How to design and implement project-based science activities. *The Science Teacher* 75 (8): 29–32.

Frank, M., and A. Barzilai. 2004. Integrating alternative assessment in a project-based learning course for preservice science and technology teachers. *Assessment and Evaluation in Higher Education* 29 (1): 41–61.

Geier, R., P. C. Blumenfeld, R. W. Marx, J. S. Krajcik, B. Fishman, E. Soloway, and J. Clay-Chambers. 2008. Standardized test outcomes for students engaged in inquiry-based science curricula in the context of urban reform. *Journal of Research in Science Teaching* 45 (8): 922–939.

H. R. 1804–103rd Congress: Goals 2000: Educate America Act. 1993. In *GovTrack.us* (database of federal legislation). *www.govtrack.us/congress/bill.xpd?bill=h103-1804*

Kirkpatrick, W. H. 1918. The project method. *Teachers College Record* 19: 319–335.

Krajcik, J. S., C. Czerniak, and C. Berger. 2002. *Teaching science in elementary and middle school classrooms: A project-based approach*. 2nd ed. Boston: McGraw-Hill.

Ladewski, B. G., J. S. Krajcik, and C. L. Harvey. 1994. A middle grade science teacher's emerging understanding of project-based instruction. *The Elementary School Journal* 94 (5): 499–515.

Marx, R. W., P. C. Blumenfeld, J. S. Krajcik, B. Fishman, E. Soloway, R. Geier, and R. T. Tal. 2004. Inquiry-based science in the middle grades: Assessment of learning in urban systemic reform. *Journal of Research in Science Teaching* 41 (10): 1053–1080.

Schneider, R. M., J. Krajcik, R. W. Marx, and E. Soloway. 2002. Performance of students in project-based science classrooms on a national measure of science achievement. *Journal of Research in Science Teaching* 39 (5): 410–422.

Siegel, L., G. Dickinson, E. Hooper, and M. Daniels. 2008. Teaching algebra and geometry concepts by modeling telescope optics. *Mathematics Teacher* 101 (7): 490–497.

Slavin, R. E. 1985. An introduction to cooperative learning research. In *Learning to cooperate, cooperating to learn*, eds. R. E. Slavin, S. Sharon, S. Kagan, R. H. Lazarowitz, C. Webb, and R. Schmuck, 5–16. New York: Plenum Press.

Steinberg, A. 1997. *Real learning, real work: School-to-work as high school reform*. New York: Routledge.

Thomas, J. W., and J. R. Mergendoller. 2000. *Managing project-based learning: Principles from the field*. Paper presented at the Annual Meeting of the American Educational Research Association, New Orleans.

Toolin, R. E. 2004. Striking a balance between innovation and standards: A Study of teachers implementing project-based approaches to teaching science. *Journal of Science Education and Technology* 13 (2): 179–187.

Van Driel, J. H., D. Beijaard, and N. Verloop. 2001. Professional development and reform in science education: The role of teachers' practical knowledge. *Journal of Research in Science Teaching* 38 (2): 137–158.

Vygotsky, L. S. 1962. *Thought and language*. Cambridge, MA: MIT Press.

Wurdinger, S., J. Haar, R. Hugg, and J. Bezon. 2007. A qualitative study using project-based learning in a mainstream middle school. *Improving Schools* 10 (2): 150–161.

Zubair, S. 2008. Silent birds: Metaphorical constructions of literacy and gender identity in women's talk. *Discourse Studies* 9 (6): 766–783.

Project-Based After-School Science in New York City

Kabba E. Colley
Eduinformatics

Wesley B. Pitts
Lehman College
City University of New York

Setting

I n the past decade, several after-school programs have been set up by university-school partnerships and not-for-profit organizations to augment the work of school systems. According to Friedman (2003), these after-school programs "provide a fundamental means of promoting opportunity and equity for urban students" (p. 79).

Fadigan and Hammrich (2004) conducted a longitudinal study of the educational and career trajectories of 152 low-income, young urban women who participated in an informal science education program. Using program records, surveys, and interview data, the researchers found that 93% of the participants went on to college following graduation from high school. In addition, just over 25% of the participants pursued careers in science, mathematics, engineering and technology, or medical/health related fields. Most of the participants' reported that their educational and career decision was as a result of the supportive, safe, and nurturing environment the program provided.

Fusco (2001) studied how students in a community-based after-school program in New York City practice science and found that science learning in such a program was very relevant because "it was created from participants' concerns, interests, and experiences inside and outside of science; it was an ongoing process of researching and enacting ideas situated within the broader community" (p. 860).

In our own work (Colley and Pitts 2003), we investigated students' understanding of ecological concepts in an after-school program using a model lesson that combined discussion with a design project. Results from our investigation revealed that overall students were engaged in the lesson and learned ecological concepts in ways consistent with their own experience. However, we also found discrepancies between students' designs and their responses to ecological questions, which were attributed to their own misconceptions.

According to Marsh and Kleitman (2002), after-school programs benefit all students, particularly socioeconomically disadvantaged students who are usually not well served by the traditional curriculum. Scott-Little and colleagues (2002) also reported that after-school programs "may have a positive impact on participants, but more rigorous research designs are necessary to provide data that clearly document program effects" (p. 387).

Eccles and Templeton (2002) conducted an extensive review of the literature on after-school programs and extracurricular activities to identify the factors that facilitate positive cognitive, psychological, and social development among youths. Their findings showed that "there is a growing evidence that youth programs focused both on prevention and promotion do increase positive outcomes and decrease negative outcomes for youth. Most interestingly, some programs not explicitly focused on academic instruction, produce gains in academic achievement, school engagement, and high graduation rates" (p. 172). However, Eccles and Templeton (2002) also noted that there is a lack of theoretical framework to connect specific program characteristics to specific youth outcomes for most of the programs studied. In the conclusion of their review, the researchers pointed out the need for more theory-driven studies of after-school programs as a way to improve after school programs.

Recent studies of after-school science programs show that they benefit students in a number of ways, including improved academic achievement, positive disposition towards science, and greater propensity to pursue careers in science (Walker, Wahl, and Rivas 2005). This is particularly true for urban students or those who are traditionally underrepresented in science (Friedman 2003). In a statewide study in New Hampshire, researchers found that students who attend after-school programs regularly showed improved learning skills as well as academic achievement (Frankel, Streitburger, and Goldman 2005). However, they also found that having high-quality staff in these programs makes a difference.

In this chapter, we report on how inner city high school students learn science through partici-pation in an after-school science program called Explore! Project-based science instruction (PBSI) was used as the main method of instruction in which students had the opportunity to plan, imple-ment, and present their own scientific investigations of urban ecosystems in Queens, New York City. We were interested in investigating how an after-school science program using a project-based approach could enable students to gain understanding and use of science process skills while also meeting science education standards. Science process skills are defined here as student ability to pose and investigate research questions that are relevant to them and their communities, and then to take actions on their research findings to improve their communities. This includes but is not limited to posing research questions, developing research plans, implementing research plans (including using scientific tools and technology, data collection, analysis, and interpretation), prepare and present research findings, and take action on research findings. By science education standards, we mean the New York City Science Performance Standards (Board of Education of New York City 1997), New York State Learning Standard 4 (New York State Education Depart-ment 1996) and the National Science Education Standards Goal 3 (NRC 1996).

We used PBSI as the preferred method of instruction because findings from prior research suggest that this is best practice for helping students learn science (Moje, et al. 2001; Blumenfeld, et al. 1991; Krajcik, Czerniak, and Berger 1998). In addition, we also believe that when students carry out projects, they learn how to generate their own research questions, design their own research plans, work collaboratively to implement their research plans, and prepare reports for presentation and peer review. Learning in a project-based context allows students to focus on questions that are relevant to themselves and their own communities. In addition, it empowers students by making them take full responsibility for their own learning. Since most students in the traditional school system are not used to this type of instruction, prior to the beginning of the program, we spent a great deal of time discussing with students and their parents the requirements and expectations of project-based science teaching and learning. We clarified roles and responsibilities and were able to address misconceptions about this approach. We acknowledged that there were going to be challenges; however, we assured the students that we would be available to serve as their facilitators, mentors, and provide resources and support throughout the program.

Introduction

The Explore! Program is an after-school high school science enrichment program sponsored by the Queens Bridge to Medicine Program at the Sophie Davis School of Biomedical Education, City College of the City University of New York (CUNY). Explore! is housed in South Jamaica, Queens-New York City on the York College CUNY campus and, at the time of this study, was currently in its ninth year of operation. The program was initially set up to serve as a pipeline into the Queens Bridge to Medicine Program, Sophie Davis School of Biomedical Education, College Now Precollege Programs and other CUNY colleges.

The Queens Bridge to Medicine Program, also located on the York College campus, grew out of a CUNY precollege initiative to help senior high school students attending NYC public schools to qualify for the seven-year BS/MD program at the Sophie Davis School of Biomedical Education. Since its inception in 1986, the Bridge Program has served to increase academic preparation of high school students entering college to pursuing careers in medicine and allied health professions. During their senior year, students accepted into the Bridge Program attend their host high school in the morning and then attend a two-semester-long academic program of courses in college freshman level English, calculus, and chemistry in the afternoon. Academic performance and other achievements of students while participating in the program are submitted as part of their application package to Sophie Davis and other colleges. The materials serve to show enhanced preparation for college-level science and preprofessional heath career programs.

A primary goal of both the Bridge Program and Explore! was to help increase the number of underrepresented minority students in NYC public high schools considering careers in medicine and allied health professions. This was focused particularly for primary career medicine at Sophie Davis. Explore! was established as a pipeline entry point to help meet this goal by recruiting and engaging students early in their high school career. Initially, Explore! was established as a Saturday out-of-school science enrichment program for students grades 9 through 11. With the help of community partners and a combination of funding from the New York City Council, the Queens Borough President's Office, and the New York State Assembly, special sessions in the summer, such as the environment science sessions we will describe, were added.

Recruitment and administration of Explore! relied heavily on the well-established ties the Bridge Program fostered with partnership high schools, college programs, and community organizations. Recruitment usually took place in the fall, with enrichment sessions in the spring and summer. Scheduling recruitment in the fall gave program recruiters the opportunity for potential participants to establish a track record of academic and career interest with their science teachers, who were asked to identify and prescreen students for the program. It also gave students the opportunity to learn about the science enrichment program and the necessary academic performance needed to apply early in the school year. A program representative visited freshman-, sophomore-, and junior-level science classes in high schools in Queens and other boroughs to attend recruitment sessions for both programs simultaneously. What was unique about these recruitment sessions was that students were exposed to health career counseling and simultaneously offered the opportunity to engage in the health career pipeline. Similarly, recruitment was conducted at college fairs, where students and their parents often came to seek information about the Sophie Davis Program and science enrichment opportunities.

Students who earned high school averages between 75 and 90 and who expressed a strong interest in science were primarily targets for recruitment. Many partnership schools usually wanted the program to focus on recruiting students with academic averages above 90. It was important, however, to engage students with averages in the lower-qualifying range because it was felt by program administrators (and the coauthors of this chapter) that with extra enrichment, these students could increase their academic performance in their host high school. In this manner Explore! was strategically placed to increase the number of high school students in the science and health careers pipeline by providing a combination of ongoing science enrichment and health career counseling.

Explore! is typically organized around several grade-level specific science enrichment themes. During the spring and summer sessions, one cohort of 25 students from each high school grade level (except for seniors) was identified as hosts. Seniors were recruited and hosted by the Bridge Program. Students in each cohort were hosted for five to eight Saturdays, usually from 9 a.m. to 12 noon, depending on the duration of the curriculum established for each theme. It was also important to establish a community- or university-based partnership that would contribute professional expertise associated with the theme and would also be able to donate personnel time and supplies to help support the delivery of the curriculum. For example, in 2005 a five-Saturday forensic science enrichment session was created for 10th-grade students. John Jay College, CUNY was the partner for this enrichment session. In this component, students spent three meetings learning ways to use different types of scientific evidence to draw meaningful conclusions from simulated scenarios in forensic science at York College. The remaining two Saturdays were hosted at John Jay College, where students were able to conduct additional laboratory exercises and learn about prerequisites to and career paths in forensic science. It is also important to note that other community relationships were established, such as partnering with a local church elders ministry to cohost an annual college fair, which became a valuable component of the program.

Methodology

This study took place in summer 2004 as part of a two-week environmental science program. The program was set up in collaboration with the Aquatic Voyagers Scuba Club (AVSC), a chapter of the National Association of Black Scuba Divers, and the Borough of Manhattan Community College (BMCC). The summer program was staffed by university faculty and members of AVSC and organized into six phases of activities that supported the environmental science theme. Students were hosted for 11 days (50-hours total) of contact time, generally between 9:00 AM. and 2:00 PM. This included two consecutive weekends of activities. Students were able to visit and learn more about four colleges that were diverse in setting and career offerings. Wet-labs and workshops were conducted on each campus. One of the most fruitful aspects of the program was the ability to engage the scientific interests of students while infusing serious college and career counseling.

The first week of the program focused on park ecology with fieldwork at Brookville Park, Queens and at York College. Each student was provided with a Student's Guide, which covered the following topics: introduction to park ecology; procedure; project planning worksheet; how to describe a terrestrial/aquatic system; how to select a site; how to classify the plants at your site; how to describe the plants at your site; soil characteristics and types of tests you can do; an example of a data sheet for describing the soil at your site; an example of a data sheet for recording soil samples; how to describe the water at your site; an example of a data sheet for recording water samples; template for research project report; glossary of terms and concepts.

In addition to the Student's Guide, there was also a Teacher's Guide for the facilitator and support staff. The Teacher's Guide covered clearly stated learning objectives; a list of New York City Science Performance Standards; list of resources (materials, tools, and technologies); instructional strategies including step-by-step description of how to organize and manage student work groups; assessment procedures; park ecology concepts; science-process skills; equity issues; and a list of references. Students collected soil, water, plant, and animal samples from Brookville Park and conducted tests on soil samples (for nitrogen, phosphorus, potassium, and pH) and water samples (for color, pH, temperature, chlorine, dissolve oxygen, nitrates, turbidity, and fecal coliform). They used computers with internet access to record their data and prepare reports. Students also had the opportunity to participate in a two-day environmental chemistry workshop at BMCC, which involved two lectures, case studies, and computer-based modeling. Students ended the first week by participating in a one-day "Discover Scuba" session sponsored by AVSC.

In the second week, AVSC sponsored a visit to the Marine Science Research Center at Stony Brook University, where students learned about research to protect the coastal environment of Long Island and NYC. The students accompanied researchers on the Seawolf, the Stony Brook University laboratory at sea. The summer program culminated with a weekend trip to Project Oceanology at the University of Connecticut, Avery Point. All participants were issued certificates of completion at the closing ceremony.

The guiding study questions were (1) How do inner city students learn science in a project-based, after school science program? (2) What science education standards do students achieve when they participate in a project-based after-school science program? (3) What are the benefits and limitations of such a program? The method of study could be described as a quasi-experimental, using a pre- and posttest treatment, without randomization or control group. This method was chosen because it was not possible to assign students randomly to the program and furthermore, finding a similar group to serve as a control was difficult, given the time and context of the program. However, in order to determine the effects of treatment, we administered a pre- and post-assessment instrument as well as a post-retro assessment (Lam and Bengo 2003). A post-retro assessment is an interview process that allows students to reflect on their prior and current science content knowledge and process skills at the end of a program. When combined with pre- and post-data, the post-retro assessment provides rich data to determine the effects of a particular treatment.

Participants

Although students were recruited mainly from the public and private schools in Queens, some of the students came from other boroughs in NYC. Eighteen students were selected to participate in the program based on a competitive application process. Admission was based on the student's cumulative grade point average, Regents examination grades, personal statement, and science teacher's recommendation. All students completed one-year equivalent of the Living Environment Regents curriculum and had a GPA of 80 or above. The student body included two 9th graders, eight 10th graders, three 11th graders and five 12th graders. The female-to-male ratio was 5: 4. The race/ethnic make-up of the students was 2-Asian, 12-Black, 2-Latino, 1-White, 1-Native American. Most students admitted into the program came from low- and middle-income families, who were first- and second-generation immigrants.

Instrument and Data Collection

The pre- and post-assessments consisted of six items selected from the June 2001 New York State Regents Living Environment Examination (available at *www.nysedregents.org/LivingEnvironment/ Archive/home.html*). The six items were number 65, 40, 41, 42, 43 and 44. Item 65 consisted of four sub-items and tested students' ability to formulate a hypothesis and conceptualize an experiment, while items 40, 41, 42, 43, and 44 tested students' abilities to analyze and interpret data. Items 65, 40, 41, and 42 required constructed responses, while items 43 and 44 were multiple choice. The amount of time allowed for students to take the pre-assessment was one hour. The scoring of the pre- and post-assessment was conducted using the scoring key for the June 2001 New York State Regents Living Environment Examination (also available at *www.nysedregents. org/LivingEnvironment/Archive/home.html*).

Group project reports refer to reports that were collectively prepared by students at the end of their projects. Students were divided into five groups. Two of the groups had three members, while the rest had four members each. Each group developed a research plan, collected and analyzed data about the ecology of Brookville Park, and prepared a research report. The NYC Performance Standards for Science (Board of Education of NYC 1997) were used to evaluate group project reports. Specifically, each project report was read and the evidence of the standards coded as follows: did not meet standard = 0; partially met standard = 1; fully met standard = 2; unable to evaluate = UE.

The post-retro assessment consisted of two parts. The first part asked students to describe what new concepts, terms, and science skills they have learned from conducting their research, while the second part asked them to describe what concepts, terms, and science skills they had learned prior to conducting their research. The individual post-retro assessments were scored using a scoring rubric (See Table 2.1, p. 29). Two instructors scored the assessments independently and compared their results. For each student, a mean score was calculated based on the two independently assigned scores. It is important to note that prior to calculating a mean score for each student, scorers discussed the rationale for their scores, and where there was discrepancy, the assessment was read over again before a final score was assigned.

Data Analysis

The pre- and post-data and post-retro data were analyzed using descriptive statistics. Following that, a series of t-tests were conducted to determine if there were (1) significant gains in participants' science literacy and process skills and (2) significant differences in gain scores by gender. With regards to the group project reports, the following qualitative data analysis procedure was used: two instructors read each report for content, grammar, and organization. Then they read and coded each report using the codes described in *Instrument and Data Collection*. Finally, a Science Learning Standards Matrix (see Table 2.2, p. 30) was developed to summarize the results and determine if there was sufficient evidence to show that participants' work met the New York City Science Performance Standards.

Conclusions

From the pre- and post-assessment data, we learned that students made a very small gain in process skills (0.47 points). This gain was not significant ($t = 1.201$, $df = 16$, $p = 0.247$). In addition, we learned that students did not differ in their science process skills either by gender or grade level. Although no significant gains in students' science process skills were found in the pre- and post-assessment data, we found significant gains in students' ecology concepts and science process skills in the post-retro assessment data ($t = 2.516$, $df = 16$, $p = 0.0229$).

The fact that we found significant gains in the post-retro assessment data, but no significant gain in the pre- and post-assessment data, may raise questions about the assessment instrument used. However, it is important to note that the pre- and post-assessment instrument was based on a subset of a statewide science assessment instrument that was already determined to be valid and reliable by state assessment experts.

The point about program effectiveness cannot be determined with the snapshot data we collected in this one study. However, longitudinal data on student performance could be used to answer questions about program effectiveness. Looking at high school graduation rates, college admission records, career choices of former students who participated in the Explore! program, and anecdotal evidence such as informal conversations with parent and students, one can argue that the program has some value and benefits to its recipients.

The findings from students' group project reports also suggest that overall students met the following NYC performance standards in science:

- S5a—Asks questions about natural phenomena; objects and organisms;
- S5f—Works individually and in teams to collect and share information and ideas;
- S6a—Uses technology and tools to gather data and extend the senses;
- S7a—Represents data and results in multiple ways; and
- S7b—Uses facts to support conclusions.

Standards that students were able to meet (partially or fully depending on the group) were

- S2a-Demonstrates understanding of characteristics of organisms;
- S2c-Demonstrates understanding of organisms and environments;
- S5c-Uses evidence from reliable sources to construct explanations;
- S7c-Communicates in a form suited to the purpose and the audience; and
- S8b-Demonstrates scientific competence by completing a systematic observation (see Table 2.2, p. 30).

New York State Learning Standard 4 states, "Students will understand and apply scientific concepts, principles, and theories pertaining to the physical setting and living environment and recognize the historical development of ideas in science." Our observations of students before and after participation in this study and a review of their group project reports indicate no evidence that they achieved this standard in its entirety. However, there was evidence that they met two of the "Key Ideas" of the Standard:

- Key Idea 6: Plants and animals depend on each other and their physical environment.
- Key Idea 7: Human decisions and activities have had a profound impact on the physical and living environment.

For instance, when we reviewed the group project reports, we found the following research questions:

- What are the types of aquatic animals in Brookeville Park and how do they relate to each other?
- Are there any differences between various soil samples at different locations at the park?
- Is there a relationship between the acidity of the water, soil near the water, and soil farther away from the water?
- What are the similarities and differences between plant life along the water and plant life farther away from the water?
- Is the water chemically polluted?

We also found that the most common themes that emerged from the group reports included plant life, animal life, human impact, the physical environment, the living environment, species relationships, and interdependence.

In conducting their projects, the students went through a process that challenged their misconceptions about science and required them to think critically. They asked questions about sampling procedures, materials, and methods. They engaged in discussions about how to implement their projects, and who would be responsible for what. They collaborated and debated. They exchanged notes and ideas about their projects. They used tools and technology, collected and recorded data, interpreted data, prepared written reports, and finally, they presented their findings in the presence of their peers, parents, and other adults. During the presentations, each group took questions from the audience and while some groups struggled with their responses, others responded well. Overall, the instructors, parents, and students were all engaged in a rich discussion of park ecology and human impact on green urban spaces. Discussions involving the intersection of science, the environment, and the public concern for green urban spaces are often left to designated experts and scientists. However, in this program students assumed the leadership role for discussing and sharing information publicly about the environmental status of Brookville Park and its ecosystem. Environmental status and associated concerns about Brookville Park were exemplified across the student projects. Similarly, during the presentations, the discussion about the projects emerged, and the audience learned more about Brookville Park's health. The new understandings that unfolded led the students and members in the audience to engage in generative dialogues about actions that could be taken to mitigate human impact on urban green space, particularly Brookville Park.

In discussing evidence about students meeting science education standards, it is important to note another critical event that took place at the beginning of this study, which provided further evidence of student engagement in the scientific process. During the preprogram orientation session, a generative dialogue with parents and students about the purpose of the program and how it was going to be implemented took place. This exchange allowed parents, students, staff, and instructors to view the program as a community of learners where everyone was an equal stakeholder in the process. In order to initiate the dialogue, a brief introduction to PBSI as a teaching method was presented. The expectations, requirements, and challenges of PBSI were discussed, with emphasis on students: generating their own learning questions connected to their lives; taking responsibility for their own learning; working in collaborative groups; planning and implementing projects; and presenting a finished product for peer review. The teacher would serve as a facilitator, mentor, supporter, and principal investigator. Parents and students were involved in developing science learning goals (knowing and doing science within the context of an ecology unit), and discussed how field studies might be conducted to help meet the goals. Parents were most interested in practical ways they could support science learning and academic achievement for their children. They were also interested in learning more about their local environment. Students were most interested in learning ecology and how the program would help them get ready for the new school year. Program instructors and staff, on the other hand, were interested in finding innovative ways to engage students and parents to improve the Explore! program. In addition, they wanted to learn how best to assess students' science-process skills in a project-based after school science program. At the conclusion of the dialogue, it was agreed that students would co-construct their own science learning by generating their own questions, planning their own investigations, using materials and tools, and doing their own ecology fieldwork.

When one takes into account all the processes, changes, and interactions the students went through, one cannot help but conclude that the students were, "engaged intelligently in public discourse and debate about matters of scientific and technological concern" (NRC 1996, p. 13).

Through this study, we have learned that project-based after-school science programs benefit inner city students by providing them with the opportunity to conduct their own scientific investigations. By planning and implementing their own research projects and by working collaboratively, these students gained a better understanding and appreciation of the scientific process. In addition, they learned about the importance of their local parks not only as a source of recreation and a resource but also as a source of scientific exploration and knowledge. From this study, we learned that an after-school science program with a project-based approach has tremendous potential in helping students get hooked on science and thus achieve science education learning standards.

We learned that learning science in a context with one's own familiar surroundings and experiences makes sense to students and therefore helps them see the possibilities of their own success in understanding and using science-process skills. To understand student use of such skills, however, the teacher/researcher must look at multiple sources of evidence because different types of assessments can provide different information about student science learning.

Finally, this study shows that there are challenges to planning and implementing project-based, after-school science programs. The most obvious challenge we faced was with available time and needed resources. We found that the time allocated to the various activities such as planning, fieldwork and presentations of results was inadequate and needed to be extended. We constantly had to make adjustments as we implemented the program. Although we ordered soil and water testing kits, some of the kits did not arrive in time for the field work. Consequently, we made a decision not to test some samples because we would run out of chemicals and reagents. Another challenge was that we had limited space and could not accommodate the high demand from parents to enroll their child in the program. Regardless of the limitation, we are convinced that with proper funding and support, project-based after-school science programs such as the one described in this chapter could bridge the science achievement gap, particularly among underrepresented groups.

Table 2.1. Scoring Rubric for Post-Retro Assessment of Students

ECOLOGY LESSON RUBRIC FOR SCORING POST-RETRO ASSESSMENT Student: _____						
		RATING SCALE*				
CRITERIA		POST			RETRO	
		0	**1**	**2**	**0**	**1**	**2**
A	**Ecology Concept/Term:** clearly states at least one ecology concept/term. Concept/term is clearly described or explained.						
B	**Non-ecology Concept/Term:** clearly states at least one non-ecology concept/term. Concept/term is clearly defined or explained.						
C	**Science-Process Skill:** clearly identifies and states at least one science-process skill. Science-process skill is clearly described or explained.						
D	**Non-science Skill:** clearly identifies and states at least one non-science skill. Non-science skill is clearly described or explained.						
E	**Quality of Response:** well written, clear, specific, free from spelling errors, free from grammatical errors, creative, response includes multiple concepts/terms, science process skills and/or non-science skills.						
	TOTAL SCORES						
REMARKS:						
*Rating Scale: 0 = no evidence of meeting criteria, 1 = met the criteria partially, 2 = met the criteria fully.						

Table 2.2. A Science Learning Standard Matrix Showing Student Group Reports and How Well They Met Selected New York City Science Performance Standards

Standards

Group Reports	S2a/ S2c	S5a	S5c	S5f	S6a	S6b	S7a	S7b	S7c	S8a	S8b	Sum
A	2	2	1	2	2	0	2	2	1	0	2	16
B	2	1	1	2	2	1	2	2	1	2	1	17
C	1	2	1	2	2	1	1	2	1	0	2	15
D	2	2	1	1	1	1	2	2	1	0	1	14
E	1	2	1	2	2	UE	2	2	1	2	2	17

1. S2a. Demonstrates understanding of characteristics of organisms
2. S2c. Demonstrates understanding of organisms and environments
3. S5a. Asks questions about natural phenomena; objects and organisms
4. S5c. Uses evidence from reliable sources to construct explanations
5. S5f. Works individually and in teams to collect and share information and ideas
6. S6a. Uses technology and tools to gather data and extend the senses
7. S6b. Collects and analyzes data using concepts and techniques in Mathematics Standard 4
8. S7a. Represents data and results in multiple ways
9. S7b. Uses facts to support conclusions
10. S7c. Communicates in a form suited to the purpose and the audience
11. S8a. Demonstrates scientific competence by completing an experiment
12. S8b. Demonstrates scientific competence by completing a systematic observation

Rating Scale: Did not meet standard = 0; partially met standard = 1; fully met standard = 2; unable to evaluate = UE

Acknowledgments

The study reported in this chapter would not have been possible without the support and participation of several individuals and organizations. We wish to express our heartfelt thanks to the following: Dr. Femi S. Otulaja, Nicole Grimes, Monique Dawkins, Stephen DCosta, and parent coordinators, Ms. Carola Craig and Sandra Williams, for assisting in the planning and implementing the Explore! summer program. Special thanks goes to Judy Berhannan, Sujon Low, and the members of the Aquatic Voyagers Scuba Club (AVSC), a chapter of the National Association of Black Scuba Divers (NABS), Stony Brook University, Dr. William Wise at the School of Atmospheric and Marine Sciences, Project Oceanology at the University of Connecticut Avery Point Campus, York College and the Sophie Davis School of Biomedical Education-Queens Bridge to Medicine Program, at the City University of New York for providing the students with rich and stimulating science activities.

To our student participants, parents, staff and supporters, we are grateful for your contributions, commitment, and belief in the Explore! program. We extend our special thanks and appreciation. We have learned so much from you.

References

Black, P. 1987. *Report: National curriculum: Task group on assessment and testing.* London: Department of Education and Science.

Blumenfeld, P. C., E. Soloway, R. W. Marx, J. S. Krajcik, M. Guzdial, and A. Palincsar. 1991. Motivating project-based learning: Sustaining the doing, supporting the learning. *Educational Psychologist* 26 (3 and 4): 369–398.

Board of Education of New York City. 1997. *Performance standards for science.* New York, NY: Board of Education of New York City.

Colley, K. E., and W. B. Pitts. 2003. After-school science. *The Science Teacher* 70 (3): 55–59.

Eccles, J. S., and J. Templeton. 2002. Extracurricular and other after-school activities for youth. In *Review of research in education*, ed. W. G. Secada, 113–180. Washington, DC: American Educational Research Association.

Fadigan, K. A., and P. L. Hammrich. 2004. A longitudinal study the educational and career trajectories of female participants in an informal science education program. *Journal of Research in Science Teaching* 41 (8): 835–860.

Frankel, S. L., K. Streitburger, and G. Goldman. 2005. *After-school learning: A study of academically focused after-school programs in New Hampshire.* Portsmouth, NH: RMC Research Corporation.

Friedman, L. 2003. Promoting opportunity after school. *Educational Leadership* 60 (4): 79–82.

Krajcik, J., C. Czerniak, and C. Berger. 1998. *Teaching children science: A project-based approach.* Boston: McGraw Hill College.

Lam, T. C. M., and P. Bengo. 2003. A comparison of three retrospective self-reporting methods of measuring change in instructional practice. *American Journal of Evaluation* 24 (1): 65–80.

Moje, E. B., T. Collazo, R. Carrillo, and R. W. Marx. 2001. "Maestro, what is 'quality'?" Language, literacy, and discourse in project-based science. *Journal of Research in Science Teaching* 38 (4): 469–498.

National Research Council. 1996. *National science education standards.* Washington, DC: National Academy Press.

New York State Education Department. 1996. *Learning standards for mathematics, science and technology.* Albany, NY: Department Office for Diversity, Ethics and Access.

Polman, J. L. 2000. *Designing project based science instruction: Connecting learners through guided inquiry.* New York, NY: Teachers College Press.

Scott-Little, C., S. Hamann, and S.G. Jurs. 2002. Evaluations of after-school programs: A meta-evaluation of methodologies. *American Journal of Evaluation* 23: 387–419.

Walker, G., E. Wahl, and L. M. Rivas. 2005. *NASA and after-school programs: Connecting to the future.* New York, NY: American Museum of Natural History.

Students as Scientists:

Guidelines for Teaching Science Through Disciplinary Inquiry

Kanesa Duncan Seraphin
University of Hawaii

Erin Baumgartner
Western Oregon University

Setting

The National Science Education Standards (NSES) stress that science is an active process (students doing science) as opposed to a passive process (science is done to students). In order to promote students' ability to investigate effectively, evaluate and articulate scientific issues both within and beyond the classroom community, we advocate learning-by-doing in the context of disciplinary inquiry. Disciplinary inquiry involves learning about a discipline by engaging in the complete practice of that discipline. In this way, students become involved in the complete scientific process. They learn that scientific investigation has many dimensions, that it does not always proceed in a strict linear fashion, and that it involves both the articulation of the investigation process as well as the outcome. One essential element emphasized in disciplinary inquiry is the communication of discoveries to peers and the broader community. This practice, in particular, prepares students to engage in civil discourse about issues, to clearly communicate and defend their conclusions, and to consider and incorporate alternative points of view and additional information in their interpretations.

Teaching Science as Inquiry (TSI)

Teaching science as inquiry means using disciplinary inquiry to teach science as it is practiced. In this form of learning, content is both an end goal as well as a framework for knowledge construction. By testing principles and connections through the generation and interpretation of their own data, students can begin to understand the fundamental underpinnings of science. For example, testing and retesting ideas helps students to understand fundamental, but complex, ideas such as what constitutes a "scientific theory." Through disciplinary experience, students can discover why scientific theories are not converted directly into facts but instead into more robust explanations and eventually subsequent theories, with greater predictive and connective power (Young 1997).

Teaching disciplinary inquiry through the authentic practice of science involves students learning science as a community process. Students and teachers work together to discover and understand the natural world. By adopting the demeanor of scientists, students also learn to use a variety of knowledge acquisition tools and communication skills. These scientific demeanors parallel the scientific "habits of mind," such as integrity, diligence, fairness, curiosity, openness to new ideas, skepticism, imagination, and communication essential to scientifically literate individuals (AAAS 1990).

The incorporation of scientific demeanors into disciplinary inquiry teaching practices can be difficult, however, as it requires significant effort, practice, and attention (Hammer 1999). Successful use of disciplinary inquiry requires that teachers themselves have a good understanding of the particular scientific discipline (Wee et al. 2007). Even those teachers who have a strong understanding of science content do not always demonstrate the scientific demeanors considered necessary by science experts (Zembel-Saul et al. 2002).

This chapter describes a framework of teaching and learning developed to help overcome the difficulty in teaching the disciplinary inquiry of science. The framework comes from the Teaching Science as Inquiry (TSI) program developed at the University of Hawaii's Curriculum Research and Development Group (CRDG). TSI helps engage teachers and students as a community in the process of doing science through a supportive skills and content-based model that builds inquiry into existing teaching practice with gradual and sustained implementation of skills within the classroom (Pottenger and Berg 2006). The TSI model promotes deep understanding of science content and processes, which allows teachers to conduct authentic science successfully within their classrooms, helping students to build their own scientific habits of mind (Pottenger, Baumgartner, and Brennan 2007; Handler and Duncan 2006), and to participate in true scientific disciplinary inquiry practices (Pottenger and Berg 2006; Pottenger et al. 2007).

Description of Intervention

Nonlinear Phases of Scientific Investigation

The TSI learning model utilizes the nonlinear progression of actual scientific endeavors; new questions may be initiated at any time during an investigation. Further, interpretation of data often leads to the invention of new hypotheses or study designs. TSI provides a framework of instruction through five integrated and overlapping instructional phases (see Table 3.1, p. 36). TSI also allows students to engage in scientific thinking through multiple modes of inquiry (see Table 3.2, p. 37), thereby permitting opportunities for investigation and exploration that are both personally relevant and locally important.

In the TSI learning model, students are the primary actors, asking and answering their own questions in longer-term investigations. TSI emphasizes the use of a few essential concepts coupled with skill development and instruction of students by one another. Because of this, students' communication of what they have learned both within and beyond their classroom scientific community is an integral feature of the learning model.

The five-phase structure of TSI shares elements of other accepted learning models (e.g., Bybee 1993). The unique features of TSI include (a) its interwoven instructional phase and (b) the strong emphasis on the use of multiple modes of inquiry. In TSI, instruction is ongoing, multidirectional, and embedded throughout the learning process. This allows teachers the flexibility to lead students through a cycle of learning that is pedagogically and scientifically sound yet flexible enough to accommodate student-driven interests. More importantly, TSI inherently results in a student-centered program. Students act as peer instructors. They also encourage communication and community-building throughout the entire process. Learning is accomplished through a variety of modes, enabling opportunities for individual talents of students to shine and for their weaker areas to improve.

Phases of Disciplinary Inquiry

The general philosophy of the TSI instructional approach involves cyclical processes integrated with instruction. This cycle addresses many of the NSES recommendations for *More Emphasis* in the Science Teaching Standards as well as the Content and Inquiry Standards (NRC 1996; see alignments indicated in Table 3.1, p. 36). As in scientific investigations, the TSI instructional model begins with *initiation* when a student identifies a problem to be solved or asks a question about their surrounding environment. Initiation can be fully student driven or guided by teacher questions, demonstrations, presentations of experiences, or anomalies that help students formulate initial questions about target content.

Following initiation, students engage in the invention of a means to solve their problem or answer their questions. In the TSI learning model, invention can be the development of a hypothesis that will guide an investigation through testing, or it can be the design of an experiment, field study, or even an apparatus that will answer the questions or solve the problems. Thus students can invent both mental models and processes or physical artifacts. By engaging in this kind of thinking about a problem, students develop their inquiry skills and gain deeper understanding of scientific concepts.

Investigation is the part of the TSI learning model that involves the actual gathering of new knowledge. During investigations the models and processes created during the invention phase are used to guide the making of observations, testing of hypotheses, and collection and analysis of data. In true scientific fashion, investigation frequently leads to initiation of new questions and invention of new processes or artifacts that can be incorporated into the current investigative framework or later used to stimulate new investigations. Thus learning does not proceed in stepwise fashion; students are encouraged to move back and forth in a fluid manner between phases. Emphasis is on cyclical, logical processes rather than rigid, linear procedures.

The information gathered during the investigation requires *interpretation*. Interpretation is both a reflective, internal process and an objective external process. During internal interpretation, student researchers must take the time to evaluate the information they have gathered, make conclusions, perhaps pose alternative explanations, and document potential scientific errors that may have occurred within their study. This information is then presented for external interpretation by the classroom community and ideally even to the scientific community as well as the general public for evaluation and review.

Instruction is integrated into each part of the inquiry sequence. Instruction in the TSI model includes teacher-to-student as well as student-to-student and student-to-teacher instruction. As students generate their own knowledge through investigation, instruction from teacher to student becomes more limited, and the process of students sharing their knowledge with their community (both within and beyond the classroom) constitutes the majority of instruction. The

TSI cycle, as previously outlined is well adapted to the formal classroom setting (see Baumgartner et al. 2009 for a description of how the TSI phases of disciplinary inquiry are specifically used in a high school classroom).

Table 3.1. The Five Phases of The TSI Instructional Model as Grounded in the Nature of Scientific Inquiry to Illustrate Connections to NSES *More Emphasis* Recommendations in the Science Teaching Standards as Well as Content and Inquiry Standards (NRC 1996).

Phase	Description	Alignment to NSES
1. Initiation	• Originate Interest that results in a problematic focus • Develop a Focus for inquiry: a question, problem or need	Teaching Standards: Understanding and responding to individual student's interests, strengths, experiences, and needs
2. Invention	• Create a Testable Resolution (hypothesis) of the question, problem or need • Create a Test Design (experiment) or way to determine the workability or the degree of success of the resolution	Content and Inquiry Standards: Understanding scientific concepts and developing abilities of inquiry
3. Investigation	• Carry Out a Test according to design in the invention phase • Carry Out an Analysis according to the design in the invention phase	Content and Inquiry Standards: Activities that investigate and analyze science questions Content and Inquiry Standards: Using evidence and strategies for developing or revising an explanation and Implementing inquiry as strategies, abilities, and ideas to be learned
4. Interpretation	• Evaluate Results: researchers draw conclusions about the workability or success of testing • Evaluate Conclusions: community evaluates the conclusions of the research; discussion of alternative explanations and additional information	Content and Inquiry Standards: Applying the results of experiments to scientific arguments and explanations
5. Instruction (embedded in all phases)	• Communicate new concepts, methods, and connection within student community • Communicate new concepts, methods, and connections through pedagogic and other broadcast means to the larger public	NSES Content and Inquiry Standards: Communicating science explanations and Public communication of student ideas and work to classmates Content and Inquiry Standards: Using evidence and strategies for developing or revising an explanation and Implementing inquiry as strategies, abilities, and ideas to be learned

(Modified from Baumgartner et al. 2009.)

Multiple Modes of Knowledge Acquisition

In addition to the phases of inquiry, TSI emphasizes the flexibility of science by exploring a variety of different modes of inquiry (Pottenger and Son 2005; Pottenger et al. 2007). This use of multiple modes of knowledge generation and acquisition is an important aspect of disciplinary inquiry because science is practiced in many ways (Windschitl et al. 2007) and investigating the nature of science in its various aspects supports student learning through conceptual change (Tytler 2002).

The TSI modes of inquiry are detailed in Table 3.2. These modes of inquiry help to illustrate the variety of ways in which new knowledge can be acquired and employed as well as the ways in which teachers and students (indeed, all people) can legitimately do scientific inquiry. Use of multiple inquiry modes also reflects research on the process of knowledge development. Students build knowledge when they construct ideas or arguments using evidence from a variety of sources, and they achieve conceptual change when they reconstruct their ideas in light of new knowledge or after taking part in discourse with others (Zembel-Saul et al. 2002).

Table 3.2. The Modes of Inquiry Addressed in TSI

Inquiry Modes	Description
Curiosity	Search for new knowledge in informal or spontaneous probes into the unknown or predictable in external environments
Replicative	Search for new knowledge by validating inquiry through duplication; testing the repeatability of something seen or described
Technology	Search for new knowledge in satisfaction of a need through construction, production, and testing of artifacts, systems, and techniques
Authoritative	Search for knowledge new to the seeker through discovery and evaluation of established knowledge via artifacts or expert testimony
Inductive	Search for new knowledge in data patterns and generalizable relationships in data association—a hypothesis finding process
Product Evaluation	Search for new knowledge about the capacity of products of technology to meet valuing criteria
Descriptive	Search for new knowledge through creation of accurate and adequate representation of things or events
Deductive	Search for new knowledge in logical synthesis of ideas and evidence—a hypothesis making process
Experimental	Search for new knowledge through testing predictions derived from hypotheses
Transitive	Search for new knowledge in one field by applying knowledge from another field in a novel way

The use of multiple modes is an important component of TSI, which emphasizes the incorporation of inquiry and investigation in a variety of ways—not just the traditional "hands-on" approach. Provision for multiple pathways to knowledge can give cautious teachers a toehold on teaching through disciplinary inquiry because it eliminates the perceived need for inquiry to always necessitate a full-scale experiment. For example, authoritative inquiry, the evaluation of information provided by an established source, is an excellent example of true scientific inquiry that is not hands-on. The use of multiple modes of inquiry also enables more student direction because there are multiple paths to knowledge generation. Moreover, the emphasis on inquiry modes like product evaluation illustrates how scientific thinking skills apply to the broader society, which can help connect students' science experiences to their life experiences. This emphasis also enables students to develop the critical-thinking skills necessary for evaluation of information they receive from the world around them; it promotes engagement in the transitive mode of inquiry, which involves taking knowledge from one field or learning setting and applying it to another. Such activities prepare all students to use scientific thinking to approach common daily problems relevant to their lives, whether they go on to become professional scientists or not.

By incorporating a variety of inquiry modes into an existing lesson, teachers are also able to involve students in more styles of learning (also an emphasis of NSES recommendations and good teaching practices). We suggest that teachers be explicit with students about the various ways that scientists acquire knowledge. For example, a unit might begin with authoritative inquiry, where students research current knowledge about the topic of interest. This can lead to inquiry through curiosity, where questions regarding the subject matter are developed. Finally, some of these questions may be addressed through experimental inquiry.

Science Is Really Thinking Class

Rather than presenting science as a strict, linear process (i.e., the so-called "scientific method" often presented by traditional textbooks) consisting of cookbook-style exercises that may seem irrelevant to students' lives, TSI promotes the use of local concepts to engage students. As NSES recommends (see alignment in Table 3.3), teachers and students begin where they are—both geographically and experientially—to take advantage of individual interests, experiences, and needs. Students are encouraged to accept ownership of their learning and discovery. Local resources, such as research scientists and conservation groups, are used to help motivate students and provide physical and academic resources. This molding of the curriculum to fit the expertise of other teachers and experts enhances science learning and provides real models for students.

Teaching through disciplinary inquiry allows teachers the freedom to say, "I don't know the answer to that question," while still maintaining a leadership role in the classroom. As TSI emphasizes the process of discovery, investigations for which the result is unknown to both students and teacher are encouraged for better sharing of responsibilities between students and teacher. Rather than chronicling facts, teachers are encouraged to engage students in the process of inquiry by helping them to ask scientifically oriented questions to promote learning by analysis. Students are encouraged to connect their explanations to their personal body of scientific knowledge. They develop and evaluate explanations based on evidence. As part of the process,

Table 3.3. Connections Between Disciplinary Inquiry Strategies Emphasized in TSI and NSES *More Emphasis* Recommendations for Science Teaching, Content and Inquiry, and Assessment (NRC 1996).

Disciplinary Inquiry Strategy	NSES *More Emphasis*
Using multiple modes of knowledge generation and acquisition	Content and Inquiry Standards: Learning subject matter disciplines in the context of inquiry, technology, science in personal and social perspectives, and history and nature of science
Using local concepts to engage students by beginning where they are, both geographically and experientially	Teaching Standards: Understanding and responding to individual student's interests, strengths, experiences, and needs
Using local resources (e.g. research scientists and conservation groups) to help motivate students and provide physical and academic resources	Teaching Standards: Selecting and adapting curriculum Working with other teachers to enhance the science program
Using investigations for which the result is unknown to both students and teacher	Teaching Standards: Sharing responsibility for learning with students
Helping students to ask scientifically oriented questions	Content and Inquiry Standards: Activities that investigate and analyze science questions
Treating students like researchers and giving them time to interpret their own results	Teaching Standards: Focusing on student understanding and use of scientific knowledge, ideas, and inquiry processes
Giving students the opportunity to justify and communicate proposed explanations within the classroom community and perhaps the scientific community / general public for evaluation and review	Teaching Standards: Providing opportunities for scientific discussion and debate among students
Assessing students' ability to observe, ask questions, and report on their ideas and process of thinking that has lead to investigations and new information	Teaching Standards: Continuously assessing student understanding (and involving students in the process) Assessment Standards: Students engaged in ongoing assessments of their work and that of others
Emphasizing flexibility at all levels (e.g. of content, process, and difficulty) and adaptation of assessment to fit teacher's needs	Assessment Standards: Teachers involved in the development of external assessments
Using project-based units as summative performance-based assessments that require logical thinking and organizing of ideas	Assessment Standards: Assessing scientific understanding and reasoning
Using performance-based assessments as opportunities for students to collect and share their data with scientists and the public, which requires rethinking and repackaging of knowledge	Content and Inquiry Standards: Groups of students often analyzing and synthesizing data after defending conclusions

student researchers are also given time to interpret their own results. All of this, of course, much better exemplifies the authentic processes of science.

Furthermore, as scientists, students justify and communicate proposed explanations with their colleagues. This information is then presented within the classroom community, and even to the scientific community as well as the general public, for evaluation and review. In this way, students learn the difference between a statement that simply sounds scientific and a statement that they can support with scientific evidence.

Enacting the TSI Model in the Classroom

The TSI framework is flexible so that all elements can be effectively embedded into a teachers' ongoing practice. The amount of time devoted to project-based TSI units is variable. A teacher experienced with project-based learning might weave long-term projects throughout a year's curriculum, whereas a novice might implement only one or two projects during the course of a year. Even informal educators, whose contact time limits the nature of long-term projects, can effectively use the strategies from TSI.

An Exemplary TSI Unit: The Case of the Sick Coral

"The Case of the Sick Coral: Translating Authentic Research Into a Classroom Inquiry Investigation of Ocean Literacy Principles" is a TSI-based lesson series that guides high school students in developing and testing hypotheses about the differential susceptibility of coral to bleaching (see Tice and Duncan 2009 and supporting materials at *www2.hawaii.edu/~katice/GK12/index_coral. html*). This series of lessons was developed through a partnership between high school teachers, University of Hawaii educators, and Hawaii Institute of Marine Biology researchers (Figure 3.1). We offer it as a model of how TSI can be applied to classroom content.

"The Case of the Sick Coral" is place-based, allowing students to make connections between science and their own lives. In the lesson series, which was designed for Hawaii classrooms, *initiation* begins when students read a newspaper article about a recent coral bleaching event in the Northwestern Hawaiian Islands. Students are then charged with designing and implementing a study to investigate the causes of the bleaching event for the National Marine Fisheries Service. (This locally relevant "hook" can be adapted for classrooms in other parts of the world by providing students with a case study closer to their own home. For example, a group of students in the Pacific Northwest might read about salmon die-offs.)

Students are introduced to the concept of coral bleaching, and through a discussion on coral biology and physiology are led to the *invention* stage of scientific inquiry. At this stage, students develop hypotheses to explain why some corals on a given reef might bleach, while neighboring corals remain healthy. Several factors might contribute to these differences in bleaching susceptibility—different corals may have different types of zooxanthellae (some of which are more resistant to bleaching); corals may experience different microenvironmental conditions; or there may be genetic differences between corals that make them differently susceptible to bleaching. Again, this hypothesis formation activity could be adapted to local situations in other areas depending on the problem of interest.

Continuing the invention stage, students are charged with designing an experiment to determine which of their proposed explanations of coral bleaching is most likely to be accurate. To

Figure 3.1. Scientist and teacher work together to develop, through product evaluation, methods for sampling and mapping a coral reef (top). Scientist and teacher work together to experimentally sample and map a coral reef (middle). Students replicate the study using a mock coral reef in the classroom, performing the same investigative steps as the scientists and teachers (bottom).

make the activity manageable, the zooxanthellae hypothesis is excluded, citing monetary restraints (which is reflective of an actual dilemma faced by researchers). At this stage, students are presented with the coral reef they have been charged with studying. The reef is simulated in the class-room, using broccoli florets to represent healthy coral colonies and cauli-flower florets to represent bleached coral (see Figure 3.1). Students are shown the reef, and they are asked to develop a sampling protocol to determine whether genetics or envi-ronmental factors best explain the observed bleaching patterns.

In the *investigation* stage, students divide into research teams to map the location and collect a tissue sample for genetic analysis from every colony on the reef according to their sampling protocol. In addition, mock data loggers placed on the reef provide depth and temperature information.

Each coral has an identifying number, and associated with each number is a specific genotype. When the class convenes to compile their data, the teacher presents the students with the genotype information specific to each coral colony's identification number.

With environmental data, genetic data, and locations of all the coral colonies in hand, students can then address their hypotheses regarding the cause of the patchy nature of coral bleaching. In their *interpretation,* students examine the genotype frequencies of the healthy and bleached corals to test for genetic differences between the two groups, and they create a map of the reef using Microsoft Excel to search for any correlations between microenvironmental conditions and bleaching prevalence on the reef.

As students begin compiling and examining their data, they are active participants in the *instruction* phase of inquiry. Students must reconcile the mass of data they have collected and weigh the evidence in support of the competing hypotheses. To do so, they must describe their activities and explain their reasoning to one another. The instruction phase culminates with formal research reports, in which students submit their findings in the form of a scientific report or a science article in the popular press. Having the students choose a specific audience—such as a local newspaper, a community meeting, an airline magazine, a TV news station, or a scientific journal—for their report realistically replicates the activities of research scientists and provides a range of report formats.

Additional TSI-Based Lesson Ideas

Effective TSI-based lessons or lesson units involve students in investigations of problems that promote findings from specific topics as well as the desire to learn about and debate general scientific topics. These investigations contribute to learning units that are authentic, aligned to instructional goals, appropriate to the learning level of students, and multidisciplinary. Because they tend to be more student-driven, flexibility and clear communication are also necessary features.

Local issues like watershed or bird monitoring that have outcomes or products of real scientific use, can inspire authentic TSI-based lessons. Community partnerships are another authentic way to provide opportunities for students to discuss their scientific thinking with the broader community (see Baumgartner et al. 2006 for suggestions on forming student-scientist-partnerships). For example, one TSI-based project in Hawaii was built around a hammerhead shark tagging and growth project in Hawaii that ultimately involved multiple teachers and hundreds of student researchers in collecting data, which was later published in scientific journals (see Figure 3.2, Handler and Duncan 2006).

Partnerships with professional researchers are not a precondition for authentic scientific experiences. Students can use techniques, tools, and strategies employed by professional scientists to look at their local environment. For example, on their school campus (or even at home), students can map the vegetation, conduct observations or experiments to monitor populations of migratory or nesting birds, monitor for invasive or pest species, or participate in testing and monitoring the water quality in local watersheds (lesson plans and ideas for these and other activities are collected at *www.hawaii.edu/gk-12/evolution*). Teachers can also construct scenarios or "mysteries" for students to investigate (for forensics-style mysteries to teach basic biological, physical, or

Figure 3.2. Student researcher prepares to weigh, measure, and tag a juvenile hammerhead shark (*Sphyrna lewini,* top). The shark's electroreceptive organs, the ampullae of Lorenzini, are clearly visible (middle). The student releases the shark to swim away with a tag in its dorsal fin (bottom).

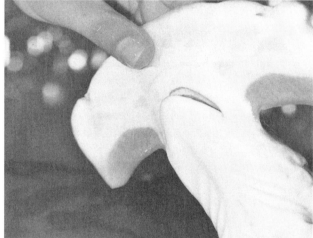

PHOTO COURTESY OF CHIS LOWE

chemical concepts, see Duncan and Daly-Engel 2006).

TSI-based units are also communicative, meaning that research ideas and results are shared among all participants and the general public in such a way that the human element of science is explored. This emphasis on science as a community of practitioners is an important aspect of teaching science as a discipline and can have real impact, not only on student learning but also on the scientific community. For example, the AntWatch project (*www.hawaii. edu/ant*) has been used to map the location of the invasive little fire ant, *Wasmannia auropunctata* in Hawaii. Over 500 students participated on the project, discovering multiple new incipient populations of *Wasmannia* as well as the first new record of another invasive species and a third species entirely new to science (Gruner, Heu, and Chun 2003). Another project, Our Project in Hawaii's Intertidal (OPIHI) has involved over 15 teachers and nearly 1,000 students in the monitoring of intertidal organisms in Hawaii (Baumgartner et al. 2009, *www.hawaii.edu/gk-12/opihi*). These projects are specific to local issues, but the principles are transferable across environments and across disciplines. In fact, the OPIHI project is similar to the California-based intertidal monitoring project known as LiMPETS (Pearse, Osborn, and Roe 2003).

Teachers can involve students in sharing their results locally by hosting a class or schoolwide symposium. The internet can be used to reach a larger audience and connect with other schools, parents, and the community. A middle school might, for example, involve all students and teachers of one grade level in a watershed monitoring program. The participation of the entire grade level team would enable teachers and students to visit multiple sites in the same watershed on the same day to collect and test water samples. The data collected could then be compiled by the students in a schoolwide research symposium and shared with the larger community on the school website. The involvement of all grade level instructors means that no one has to be inconvenienced by missing a class for the field trip, and that the teachers have an opportunity to work together to build a multidisciplinary unit with mathematics, art, reading, and writing around the common watershed theme.

Assessment

While the project-based approach fostered by TSI naturally lends itself to a capstone activity and summative assessment, there are also multiple opportunities for ongoing, formative assessment that is embedded in the lesson sequence. Assessment of TSI units can occur in a variety of ongoing and flexible ways; students observe, they ask questions, and they report on their ideas and their processes of thinking that has lead to investigations and new information. In "The Case of the Sick Coral," for example, student groups first work to construct a poster list of possible causes for coral bleaching and then a method for studying the mock reef. As a group, the class discusses and agrees on a sampling method. Assessment can also occur in the field, as with OPIHI where student knowledge is naturally assessed by their abilities to employ sampling techniques correctly and to identify organisms they encounter. Students participating in watershed monitoring projects, like the seventh-grade teams described above, have a natural assessment opportunity when the various teams return to school to pool the data from the different sites. If each team has not employed the proper data collection techniques, it quickly becomes obvious as they present their results.

TSI emphasizes flexibility at all levels, and teachers are encouraged to adapt assessment strategies to fit their needs. The TSI structure is accessible to teachers across a broad spectrum of learning levels and content areas. Because content is flexible, assessment can address the content targeted by the individual teacher in traditional ways (including written tests, quizzes, oral questions, and worksheets) as well as nontraditional ways (such as posters, essays, poems, oral stories, and the demonstrated ability to participate effectively in projects).

TSI lends itself particularly well to summative performance-based assessments, which require logical thinking and organizing of ideas. Typical end-of-project assessments seen in TSI lessons include formal presentations with PowerPoint, scientific reports, forensics "legal reports," pamphlets, town hall meetings (i.e., classroom debates), science posters, and open houses. Such performance-based assessments provide natural opportunities for students to collect and share their data with their school community, with scientists, and with the general public. In one TSI-based project, for example, students who had been investigating the success of shearwater chicks on offshore islets in Kailua Bay, Hawaii developed signage for visitors to a local beach park to share information about the nesting birds and tips to avoid harming the birds.

Students in TSI-based projects also communicate the knowledge they have learned beyond their classroom in informal ways. In the Hammerhead Shark Tagging and Growth Study, one student wrote, "At first I thought sharks were mean and vicious, but now my perspective has changed a whole lot…. I will successfully take this knowledge and pass it on to my fellow friends and family." Another student wrote, "To tell you the truth, I was able to also teach my brother that sharks are not really all scary" (Duncan 2010). These examples showcase the ability of TSI-based projects to empower students to communicate and share their scientific discoveries.

Teachers can assess these often hidden gains in students' knowledge, as well as positive shifts in attitude, by asking students opinions about the learning unit and by asking how the students perceive their own knowledge and attitude to have changed. This student self-reflection also provides explicit awareness of the inquiry phases and modes, enabling students to consciously learn the process of doing science. For example, at the conclusion of "The Case of the Sick Coral" lesson series, teachers ask students to rate their ability to apply knowledge from the coral genetics and mapping work to a problem of plants around their school. Students are first asked to score their confidence in doing this type of work, and then they are asked to detail some of the processes they might use and some of the factors that might need to be considered. In another example, students in the Hammerhead Shark Tagging and Growth Study conducted a mock tagging experiment as a form of assessment. Each group was given a picture of a different organism for which they had to design a tagging research program based on a research question they were interested in and using supplies available in the classroom "tagging kit," consisting of some real tagging equipment and some photographs of equipment. As a teacher participant observed after watching her students conduct this mock tagging experiment, "I really like this. They [the students] are actually THINKING [teacher's emphasis]. It is really hard to get them to think for themselves" (Duncan 2010). This type of assessment addresses students' attitudes, feelings of competency, and ability to transfer knowledge and skills to a novel situation (the transitive mode of inquiry). Such transfer of skills is the essence of a persons' ability to conduct scientific observations and experiments. It is the hallmark of the TSI framework.

Directions for Future Work: Professional Development in TSI

Professional development in disciplinary inquiry can help teachers become successful facilitators of scientific inquiry, enabling them to create classrooms that function as a community of scientists—where students learn science by engaging in the practice of science. The professional development programs that we are establishing teach educators to facilitate students' scientific literacy by helping students understand not only basic scientific concepts, but also the process that has been used to gain and refine those concepts over time (Figure 3.3, p. 47). Through the TSI program, teachers learn to develop authentic science activities that rely on and emphasize the use of tools and techniques that are also used by professional scientists. Teachers learn to help students evaluate and decide which tools and techniques to use, and teachers are encouraged to provide students the opportunity for social interactions within the context of science inside the classroom and beyond. When teachers teach disciplinary inquiry, they are effectively guiding students' thinking and reasoning through the judicious use of discussion, insight, and assistance—thereby teaching science *as* and *through* inquiry rather than *by* inquiry (see van Zee et al. 2005).

The TSI professional development institutes that we have facilitated were centered on specific content foci in order to provide teachers the most authentic experience possible. For example, theme topics for TSI institutes have included Astronomy (elementary), Aquatic Science (high school), Sustainable Energy (middle school), Physics and Physiology (middle to high school), and Why Things Sink and Float (middle school). (See Table 3.4, p. 48 for a sample instructional sequence for an aquatic science TSI.) In addition to creating new content foci for TSI, our future work will include the development of modularized TSI trainings that guide teachers through a series of institutes related to a central theme.

Our current TSI professional development structure consists of a two-day (17-hour) workshop combined with a three-hour follow-up experience. The purpose of the initial workshop is to ground teachers in the TSI instructional model through a range of topic-specific content. Participants are immersed in the phases and modes of TSI in a series of activities through an iterative process during which each aspect of instructional practice is introduced in an activity. Participants form a strong foundation in disciplinary inquiry while at the same time becoming part of a network of teachers and facilitators to interact with in support of their science inquiry teaching practice.

Three to four months after the initial workshop, teachers participate in a follow-up experience (three hours long) to help them maintain the professional community developed during the institute and to provide further support for teachers. The follow-up also builds accountability by providing a venue for teachers to share the results of their implementation with their colleagues. In addition, this aspect provides opportunities for continued development as participants share with one another which inquiry strategies and which science lessons have (and have not) worked for them. During the follow-up, teachers share their reflections on their attempts to incorporate inquiry, participate in a structured peer-feedback session, participate in Q&A with facilitators, and prepare a lesson plan to share with colleagues for critique.

In the past two years, we have developed and conducted TSI professional development institutes ($N = 18$) on the islands of Kauai, Hawaii, Maui, and Oahu, as well as in Guam and China. An external evaluation conducted by the Pacific Resources for Education and Learning (PREL) in 2008 indicated positive participant feedback. Ratings on the General Workshop Evaluation Survey, which PREL used to address participants' opinions of the effectiveness of the PD, were high. The respectfulness of the learning climate (the community and collaborative nature of TSI) rated highest overall.

Teachers' written comments for their PD accreditation portfolios also suggested positive influence of the PD on the implementation of TSI strategies into existing classroom practice. For example, a high school teacher on the island of Hawaii, stated, "The workshop has helped me immensely in my curriculum planning this year. Everyday I see examples of how this type of learning is effective and engages the students in what they're learning." A teacher of seventh-grade students on the island of Kauai stated similar sentiments about the TSI workshop in her PD

Figure 3.3. Teachers participating in a TSI physical aquatic science professional development course begin the process of understanding wave dynamics by inquiring into the effect of shoreline incline on wave shape (top). Teachers then empirically derive formulas for calculating various wave properties such as wave height, length, and period (middle). Finally, teachers apply this knowledge to the study of a real beach (bottom).

accreditation portfolio: "One of the most important things I gained from this workshop was a new perspective on teaching Scientific Inquiry. There have been many previous occasions where I had a really great idea for an activity or project, but then scrapped the idea because it didn't really fit the traditional 'scientific method' model. I now understand that scientific inquiry does not always have to be a long, drawn out process of specific steps. After attending the TSI workshop, I have tried to incorporate this newfound idea into my own lessons on a regular basis!"

Table 3.4. Instructional Sequence for TSI: Aquatic Science.

Instructional Component	Content/Skills
Introduction: Observations of an unknown marine object	**What is a scientist?** Open-ended discussion on the professional and personal qualities of scientists.
Activity: Thought swap—Ocean Literacy Essential Principles (OLEP)	**Science concepts in teaching aquatic science**—Discussion of OLEP and relation to content standards (local and national).
Professional practice reflection: Scientific habits of mind and OLEP	**What is scientific literacy?** How does ocean literacy promote and relate to scientific literacy and how can we support its development in students? – Discussion of human element of scientific research and interpretation of scientific habits of mind.
Activity: Practicing scientific habits of mind	**Fish printing and fish anatomy**—Teachers use scientific habits (e.g., curiosity and observation) to learn about fish.
Professional practice reflection: The phases of inquiry	**Inquiry instructional sequence**—Teachers are guided through CRDG's phases of inquiry: Initiation, Invention, Investigation, Interpretation, and Instruction.
Activity: Practicing scientific habits of mind	**Observation and critical thinking**—Teachers are guided through a material safety exercise about the hazards of the "dangerous chemical" dihydrogen monoxide (H_2O).
Professional practice reflection: Scientific habits of mind	**Modes of inquiry**—Teachers explore the various ways that the demeanors of science are important in everyday life and how these demeanors can play out in daily instruction.
Activity: Working through the phases of inquiry using sand as a medium for exploration	**Sand investigation**—Teachers investigate properties of sand and learn about sand formation, then apply their knowledge to match local "unknown" samples with the appropriate beach. Finally, they compare their samples with samples from around the world.
Professional practice reflection: The modes of inquiry	**Modes of inquiry**—Teachers explore the various ways scientists go about their work through different modes of inquiry and apply the modes to their investigation of sand.
Activity: Combining phases and modes of inquiry into a complete lesson	**Beach pollution**—Teachers work together to initiate, design, and conduct a study of pollution at a nearby beach (located at an ocean, lake, stream, or river).
Professional practice reflection: Application of inquiry	**Reflection and development of inquiry activity**—Using their existing curriculum, teachers redesign an existing lesson to be more inquiry-based with support from facilitator and their peers.

chapter 3

Table 3.4. (continued)

Implementation of strategies in classroom: With support from online resources and online learning community	Extended practice—Teachers try inquiry strategies with their students, examine impact on self and students, identify needs, challenges, and successes. Teachers are supported by peers and facilitators who "coach" online.
Follow-up training: Three-hour, face-to-face group training with TSI teacher cohort, including a series of three activities investigating density in aquatic science.	Density stations, TSI review, support, and accountability—Teachers investigate density in three different activities (showing the flexibility of the TSI pedagogical structure to teach a concept through multiple modes) followed by discussion and debriefing with fellow participants about their own lessons and their implementation of inquiry strategies in their classrooms.

Summary

TSI provides a foundation for engaging both teachers and students in disciplinary inquiry. The use of a flexible and fluid learning cycle where student instruction and communication are interwoven with multiple modes of knowledge generation is the hallmark of the TSI approach. This learning model provides a way for learners to gain scientific literacy by engaging in the full and total practice of science, thus building the demeanors and habits of mind associated with practicing scientists. Because the model is so flexible, it can be easily incorporated at any level and with a variety of content. It is aligned with the NSES *More Emphasis* recommendations for science teaching and has been used successfully in a variety of venues. The flexibility and applicability of TSI has enabled its use with programs sponsored through the National Science Foundation STEM Graduate students in K–12 Teaching (GK–12) program, Sea Grant, the Centers for Ocean Science Education Excellence (COSEE) network, the Berniece P. Bishop Museum, the Kohala Center, and the Maui Economic Development Board. Recent use of the TSI program as a framework for preservice teachers in a biology methods course at Western Oregon University also shows promise for TSI as a mechanism for preservice teacher professional development.

References

American Association for the Advancement of Science (AAAS). 1990. *Science for all Americans*. New York, NY: Oxford University Press.

Baumgartner, E., K. D. Duncan, and A. T. Handler. 2006. Student-scientist partnerships at work in Hawai'i. *Journal of Natural Resources and Life Sciences Education* 35: 72–78.

Baumgartner, E., C. J. Zabin, J. K. Philippoff, E. Cox, and M. L. Knope. 2009. Ecological monitoring provides a thematic foundation for student inquiry. In *Inquiry: The key to exemplary science*, ed. R. Yager, 191–209. Arlington, VA: NSTA Press.

Bybee, R. 1993. *Reforming science education: Social perspectives and personal reflections*. New York: Teachers College.

Duncan, K. M. 2010. A partnership approach to improving student attitudes about sharks and scientists. *School Science and Mathematics* 110 (5): 35–47.

Duncan, K. M., and T. S. Daly-Engel. 2006. Using forensic science problems as teaching tools: Helping students think like scientists about authentic problems. *The Science Teacher* 73 (11): 38–43.

Gruner D. S., R. A. Heu, and M. Chun. 2003. Two ant species (Hymenoptera: Formicidae) new to the Hawaiian Islands. *Bishop Museum Occasional Papers* 74: 35–40.

Hammer, D. 1999. *Teacher inquiry*. Center for the Development of Teaching Paper Series. Report: ED433997.

Handler, A. T., and K. M. Duncan. 2006 Hammerhead shark research immersion program: Experiential learning leads to lasting educational benefits. *Journal of Science Education and Technology*. 15(1): 9–16.

National Research Council (NRC). 1996. *National science education standards*. Washington DC: National Academy Press.

Pearse, J., D. Osborn, and C. Roe. 2003 Long-term monitoring program and experiential training for students (LiMPETS): Monitoring the sanctuary's rocky intertidal with high school students and other volunteers. In *Ecosystem Observations*, ed. J. Caress, 6–9. Monterey, CA: Monterey Bay National Marine Sanctuary.

Pottenger, F. M., and K. Berg. 2006. Inservice training in science inquiry. Paper presented at the Pacific Circle Consortium Conference. Mexico City, Mexico.

Pottenger, F. M., and Y. Son. 2005. Inquiry, Socratic questioning, and student reflective assessment in the science classroom. Paper presented at the Pacific Circle Consortium Conference. Sydney, Australia.

Pottenger, F. M., E. Baumgartner, and C. A. Brennan. 2007. Teaching science as inquiry (TSI): Teacher professional development. Paper presented at Hawaii Charter School Administrative Office Breaking Barriers to Learning Professional Development Conference. Honolulu, HI.

Tice, K. A., and K. M. Duncan. 2009. The case of the sick coral: Translating authentic research into a classroom inquiry investigation of ocean literacy principles. *Current* 25 (2): 1–9.

Tytler, R. 2002. Teaching for understanding in science: Constructivist/conceptual change teaching approaches. *Australian Science Teachers' Journal* 48 (4): 30–35.

van Zee, E. H., D. Hammer, M. Bell, P. Roy, and J. Peter. 2005. Learning and teaching science as inquiry: A case study of elementary school teachers' investigations of light. *Science Education* 89 (6): 1007–1042.

Wee, B., D. Shephardson, J. Fast, and J. Harbor. 2007. Teaching and learning about inquiry: Insights and challenges in professional development. *Journal of Science Teacher Education* 18 (1): 63–89.

Windschitl, M., K. Dvornich, A. E. Ryken, M. Tudor, and G. Koehler. 2007. A comparative model of field investigations: Aligning school science inquiry with the practices of contemporary science. *School Science and Mathematics* 107 (1): 382–390.

Young, D. B. 1997. Science as inquiry. In *Envisioning process as content: Towards a renaissance curriculum*, eds. A. Costa and A. Liebmann, 120–139. Thousand Oaks, CA: Corwin Press.

Zembel-Saul, C., D. Munford, B. Crawford, P. Friedrichsen, and S. Land. 2002. Scaffolding preservice science teachers' evidence-based arguments during an investigation of natural selection. *Research in Science Education* 32: 437–463.

"Who Ate Our Corn?"

We Want to Know and so Should You!

Craig Wilson and Timothy P. Scott
Texas A&M University

Juan D. López, Jr.
USDA-ARS, Areawide Pest Management Research Unit

Setting

It seems that children are born with innate curiosity. Small children are renowned for always asking questions to the point of distraction. "What's this?" sounds like a mantra as they touch, smell, taste, listen, and look at the exciting world around them. Children are consummate observers. In short, we all enter the world with the tools of scientists but, somehow, somewhere, that thirst to question is squelched in classrooms where there is insufficient time for questions or to question. Memorization appears to be prized ahead of questioning. The Future Scientists Program is built around a question, but with the larger goal of reawakening the scientist in all of us. Society will be better served by a scientifically literate population, not only able to explore science, but to be in a position to question it.

Introduction

"Who ate our corn?" sounds like a simple enough question but is designed to intrigue students and launch them on a quest to find answers and to ask more questions. That is the goal of the Future Scientists Program, which is funded by the USDA/Agricultural Research Service (ARS) under the auspices of the USDA/Hispanic Serving Institutions National Program (HSINP) and administered through the Center for Mathematics and Science Education (CMSE) within the College of Science at Texas A&M University (TAMU). It establishes a collaborative community of teachers, students, parents, and research scientists with the underlying theme that, "Everyone deserves to share in the excitement and personal fulfillment that can come from understanding and learning about the natural world" (NRC 1996, p. 1). Importantly, it is funded by USDA/ ARS so that cash-strapped schools and districts are not burdened financially and thus a major impediment to participation by enthusiastic teachers is removed (Alberts 2008).

Participation by teachers at a two-day mini-summer institute at their local USDA/ARS research laboratory prepares them to help their students to: "Engage intelligently in public discourse and debate matters of scientific and technological concern" as recommended by the National Science Education Standards (NSES) Goal 3 (NRC 1996, p. 13). Initial involvement of teachers ensures sustainability to the program over the period that the teachers are in the

Figure 4.1. Students at Bonham Elementary School in Bryan, Texas, display various stages in the life of the corn earworm, having presented their research earlier.

classroom. The focus is on the billion dollar annual cost of damage caused by and efforts to control the agricultural pest, the corn earworm, *Helicoverpa zea* (Boddie). The Future Scientists Program becomes the launching pad for students in grades K–12 to develop the abilities necessary to do scientific inquiry and to develop understandings about scientific inquiry that allows them to conduct research projects studying the worm. Guided by their teachers and in consultation with program staff and USDA/ARS scientists, students grow the insect through its life cycle and begin to inquire and to research ways to lessen the adverse impacts of the worm feeding on the developing corn ear. At the end of the school year, the students present their research findings in a scientific forum at their local USDA/ARS lab. This is known as the Student Research Presentation Day (SRPD) where posters are displayed and where the students give power point presentations of their research before an audience of their peers, their teachers, parents, research scientists, and invited members of the community. It is an alternative assessment strategy as recommended by Scott et al. (2008) that can "take the focus off grades, put more emphasis on learning with understanding, and reduce test anxiety and stress on struggling learners" (p.13) as the project is all-inclusive and not focused on science-oriented students. Teachers are encouraged to select students to present who have been most engaged and involved in the project or those they feel have benefited most by participation. The philosophy is similar to that expressed by Don Rea, a retired research chemist and a research director at NASA who volunteers in a program of the American Association for the Advancement of Science (AAAS). He stated, "We are not interested in providing tutoring to students who want to become scientists. Our objective is to try to increase science understanding of the entire class" (Lempinen 2006, p. 1892). The citizenry of the United States needs to reconnect and to reengage with science to be able to make informed decisions.

The program received the 2008 USDA/ARS Administrator's Outreach, Diversity, and Equal Opportunity Award for supervisory/managerial personnel for its efforts to help increase student involvement and interest in science, as well as to spread the word about current ARS research. Since winning the award, the program is expanding nationwide starting with Arizona and California in 2009–2010 as there are over 100 USDA/ARS laboratories nationwide. The USDA has begun discussions with the CMSE to use the Future Scientists Program as a national model for their agency. It would serve as a vehicle to illustrate basic scientific concepts and the societal benefits of local agricultural research, with the hope of inspiring and supporting the next generation of scientists and of broadening the knowledge base of all students and eventual citizens who choose to follow alternative career paths. This is a win-win situation, where the partnership is beneficial to the educational and local community as well as to the participating scientists and to the Federal Agency (Beck et al. 2006).

Aspects of the NSES Visions

The program addresses each of the goals of school science that underlie the NSES. The focus of this monograph is on goal 3 ("Engage intelligently in public discourse and debate about matters of scientific and technological concern") so the alignment of the project to that goal will be explained first, followed by its relationship to the other three goals, since the four are interrelated.

"Science works best in a culture that welcomes challenges to prevailing ideas and nurtures the potential of all people. Scientific ways of thinking and of reevaluating one's views in light of new evidence help to strengthen a democracy" (Omenn 2006, p.1696). This lofty goal is one espoused by the Future Scientists Program. In order for it to be achieved, it is the teachers who are first helped to feel comfortable in a research environment. Such a situation, to many, is new and unsettling. As recommended by Pfund et al. (2006), collaborating scientists are chosen specifically by the program director for their ability to communicate their research and for the commitment they have previously shown for educational outreach. This research is of immediate relevance and interest to the teachers because each USDA/ARS laboratory has a focus on agricultural problems afflicting growers and ranchers in its vicinity. Often these problems have nationwide or worldwide implications and applications. For example, the summer institute at the USDA/ARS Carl Hayden Bee Research Center in Tucson, Arizona, introduced teachers to the current problem of Colony Collapse Disorder (CCD) that affects both the livelihood of apiarists and adversely impacts crop pollination and therefore the nation's food supply. Similarly, when teachers attend the U.S. Salinity Laboratory at USDA/ARS Riverside, California, they learn about its research mission to develop new knowledge and technology to find solutions to problems of crop production on salt-affected lands, to promote the sustainability of irrigated agriculture, and to prevent degradation of surface and groundwater resources by salts, toxic elements, pesticides, and pathogens. These are all pressing problems where knowledge and understanding are essential for citizens to be able to understand and balance the arguments of the competing interests.

The one common theme at each summer institute is the study of the corn earworm as a research model because contact with insects is something that has immediate appeal to students of all ages. The larger picture is that the USDA/ARS is pursuing alternative courses of action to control the amount of damage caused by the corn earworm through Integrated Pest Management (IPM). This is a concept of preventive suppression of a mobile insect pest species throughout its geographic range involving compatible control measures, rather than reactive field-by-field control. The mission of the Area-wide Pest Management Research Unit (APMRU) in College Station, Texas, is to develop, integrate, and evaluate multiple strategies and technologies into system approaches for management of field and food crop insect pests. It is a concrete example that the teachers and students learn about in depth and this allows them to see the larger picture. They initially study the insect's life cycle before engaging in research projects and perhaps coming to understand that their personal decisions have to be made from a foundation of knowledge and a "broad base of scientific understanding" (Texley and Wild 1996, p.27). Only then are they equipped to engage in informed debate on matters of scientific and technological concern.

Goal 1 of the NSES is to help students to, "Experience the richness and excitement of knowing about and understanding the natural world" (NRC 1996, p. 13). The program is designed to encourage science teachers, students, parents, and other members of the local community to become knowledgeable about the world-class agricultural research being undertaken by ARS scientists in their own backyard. The students are encouraged to get excited about science and to stay interested in science. Where better to start than with research that is cutting edge and relevant to the students' daily lives? Most USDA/ARS laboratories are located in rural communities where students still have an attachment to and an interest in "the land." For example, Manuel Chavez, a teacher from Mammoth, Arizona, wrote the following about his reason for joining the program: "Being from a rural school, many times our students come across insects and bugs that inhabit our environment in the San Pedro River Valley where many farms still grow corn. When the corn begins to ripen in late August and early September, families from the tri-community of Mammoth, San Manuel, and Oracle, visit the farms to purchase white corn for making masa for green corn tamales. It would be beneficial to have information to bring back to the students and involve them in the life cycle of the corn earworm since many of them encounter the insect when preparing masa." Beyond this initial engagement, students gain firsthand insight into scientific research and possible future careers, courtesy of a common agricultural pest, the corn earworm and begin to see Agriculture as something other than simply the planting and harvesting of crops by hand. Also, corn and its coevolved pest, corn earworm, truly belong to the Americas and thus present opportunities for students to explore numerous aspects of the sciences.

This insect is also known as the tomato fruitworm, cotton bollworm, sorghum headworm, or the strawberry fruitworm, depending on which crop it is consuming. The worms generally live and eat inside corn husks and develop into adults that disperse into younger corn or other crops during their relatively short life cycle (about one month long). It is precisely for this reason that the worm makes the perfect specimen or model for introduction into school science classrooms, where students and teachers can easily follow their complete progression from eggs to larvae to pupae to moths. In exchange for a small investment of time, students can reap a lifetime of lessons, scientific and otherwise. The students become part of an ongoing research program where they are encouraged to learn about the insect's life cycle through hands-on inquiry as they are all provided with their own insects to follow through their typical life cycle. Students, with guidance from their teachers, begin by developing their observation skills, pose questions, conduct research, collect data, and design experiments that examine ways to reduce the tremendous damage caused to crops by the insect. It is a "real-world" problem and the students are engaged in helping to find solutions.

Goal 2 of NSES is to "use appropriate scientific processes and principles in making personal decisions" (NRC 1996, p. 13). The Math and Science Partnership program was established as part of the No Child Left Behind Act and mandated the involvement of scientists and mathematicians in K–12 Education (Drayton and Falk 2006). The Future Scientists Program does just that but is careful to select scientists who have a passion for educational outreach and the ability to communicate successfully the content and process of their research (Wilson 2001). It is the teachers who are initially introduced to the different research projects at the lab closest to their schools. This provides better preparation for them to help their students to engage in inquiry using appropriate scientific processes and principles when making decisions about the research projects they will develop. It is these projects that Ledley et al. (2003) so eloquently describe as "enabling students to develop a sense of the sometimes graceful, yet sometimes, stumbling, esoteric, and exciting schema that scientists employ to do science" (p.92). Through this sometimes tortuous process, students become immersed in the science of the corn earworm and are helped to understand that what they do as well as how they conduct their research is vitally important if it is to be credible and of value. Similarly, they are encouraged to both ask questions and to question. They become credible partners in the process and have been found to respond positively, coming to realize that decisions they make actually do matter.

Goal 4 of the NSES is to help students to, "increase their economic productivity through the use of the knowledge, understandings, and skills of the scientifically literate person in their careers" (NRC 1996, p. 13). As has been frequently and pessimistically noted, this nation is falling behind in producing science graduates as evidenced in the report: *Rising Above the Gathering Storm* (COSEPUP 2005) where one remedy suggested is the use of summer institutes "to allow teachers to keep current with recent developments in science, mathematics, and technology and allow for the exchange of best teaching practices" (p.6). The report also hypothesizes that exposure to hands-on research will encourage students to maintain an interest in science even if they do not eventually pursue such a career in science. The Future Scientists Program is one such remedy that has perhaps been given greater credibility by the recent changes in presidential leadership that have resulted in a renewed emphasis on science education. Lederman and Malcom (2009) suggest that, "Scientists must help the adult public (and not just students) develop a clear understanding of what science is and that science education should be "a way of knowing about the world based on evidence and logical analysis" (p. 1265). They also recommend that the efforts should start locally. The Future Scientists Program not only targets science teachers and their students but also the parents of those students as well as members of the local community. For example, the USDA/ARS Watershed Management Research Unit in Tombstone, Arizona, interacts with all the communities within the Walnut Gulch Experimental Watershed, seeking input and then transferring research findings, both knowledge and technology to stakeholders, decision makers, and the public. This is just one of about 100 USDA/ARS laboratories nationwide, each charged with serving stakeholders and the public. The Future Scientists Program has the potential to develop into a national model designed to help each laboratory to become a valuable educational resource within its local school communities.

Program Description

The Future Scientists Program is a professional development opportunity for teachers that helps them create a science learning community by first linking USDA/ARS research scientists and their technicians with the teachers at a mini-summer institute held at a USDA/ARS research laboratory. It begins by working with teachers on inquiry-centered science as promoted by the NSES. These actions have encouraged many specific changes in teaching school science in ways endorsed by the National Sciences Resources Center (NSRC). This partnership has encouraged teachers to work with their students so that they "learn to ask questions, experiment, develop theories, and communicate their ideas" (p. 1). There is the potential to eventually operate this program nationwide. It has begun an initial expansion into laboratories in Arizona and California, both states that have large Hispanic populations.

Science teachers within commuting distance of their local USDA/ARS research laboratory are recruited to take part in a two-day institute. They must have the written support of their school and district and agree to participate in all aspects of the project. This includes allowing program staff to work with their students during the year for at least one day while the district has to agree to provide a substitute teacher so that the participating teacher may return with students to the research laboratory in May at the end of the school year to attend the Student Research Presentation Day. This is the only expense borne by the school. The teachers are paid a stipend for their involvement and provided with a digital microscope. About four separate institutes are held each year at four separate laboratory sites with 10 teachers at each site. For example in June 2009, there will be four Future Scientists institutes, one at each of the USDA/ARS laboratories in Tucson and Tombstone, Arizona, and at Riverside and Parlier, California. Previous states involved have been Arkansas, New Mexico, Oklahoma, and Texas.

Summer Institutes

Once at the institute, the 10 teachers learn about the research and life cycle of the corn earworm. This involves the handling of the insect in all its life stages, field work, and exposure to current USDA/ARS research concerning the area-wide pest management research unit (APMRU). The latter is a concept of preventive suppression of a mobile insect pest species throughout its geographic range, rather than reactive field-by-field control. The mission of the APMRU is to develop, integrate, and evaluate multiple strategies and technologies into system approaches for management of field and food crop insect pests. The participants then examine previous student research projects and brain-

Figure 4.2. The teachers study field corn for worm infestation rates and dig for pupae in the soil below the plants.

storm ways to answer their own questions about the insect. This enables them to become familiar with growing the corn earworm insect through its complete life cycle and to learn about the cutting edge research that is taking place nationally.

The focus is on the life cycle of the corn earworm but the program has broader scope and appeal since it touches on the Standards in science for K–4 students Life Cycles of Organisms; 5–8 Structure and Function in Living Systems/Reproduction and Heredity; 9–12 Biological Evolution and Interdependence and Behavior of Organisms. Some teachers at the high school level have been able to develop projects for Advanced Placement (AP) environmental science classes and environmental systems classes, in addition to studies offered in the regular biology classes. The activities are also ideally suited for implementation in vocational agriculture classes. In lower grades the emphasis is on scientific processes and investigation skills while helping teachers to use the reforms recommended by the content standards of life cycles.

A common theme at all Future Scientists institutes is investigation and experimentation. Actual research on the life cycle of the corn earworm is shared but, with the pressure on teachers to "teach to the test" and to focus on specific content, each Future Scientists institute is customized for each set of state standards and for each laboratory site. This exposure to the work of some of the other research units at each laboratory enables the teachers to better understand what research is taking place locally. Their interactions with scientists from different backgrounds allow the teachers to break down the stereotypical picture of scientists and to share that understanding with their students. Abruscato (2000) suggests that "stereotypes of scientists may discourage children from considering careers in science or science–related fields" (p. 16).

Year of Research

The teachers' interaction with the carefully selected research scientists (Russell, Hancock and McCullough 2007) allows them to learn about cutting-edge research firsthand and to better understand that "scientific inquiry refers to the diverse ways in which scientists study the natural world and propose explanations based on the evidence derived from their work" and that "inquiry refers to the activities of students in which they develop knowledge and understanding of scientific ideas, as well as an understanding of how scientists study the natural world" (NRC 1996, p. 23). Then the teachers have to decide how to apply elements of the research experiences when back in their classrooms with students. This is partly accomplished with guidance from the program director, and in collaboration with the other teachers as recommended in the NSTA's *Pathways to the Science Standards, High School Edition* (Texley and Wild 1996) Professional Development Standard D.

The program has as its focus "learning from work on local issues with personal relevance and local importance" because *Helicoverpa zea* is an opportunistic feeder with a national range. The moth migrates in from Mexico and its larvae feed primarily on corn crops but also cause enormous damage to other crops such as cotton, sorghum, tomatoes, and strawberries. The latter crop is particularly adversely affected in California where losses to growers rise into the hundreds of millions of dollars. The USDA/ARS labs are located, for the most part, in rural, agricultural communities where the students still have a stake in and a background in agriculture. The studies in the institutes have direct and immediate relevance for students, their families, and their communities. This is the hook to get them to engage the students in the program. They often have a vested interest in helping to solve the problem of worm damage and then become involved and immersed in the science of the corn earworm insect. Initially this involvement is with guidance from their teacher and then with help from scientists, who visit them in school, loan equipment, and are available to answer e-mailed questions. Also, the program director visits schools during the school year to support the teachers and students in their research. He encourages the students to learn about and to question the pros and cons of the worm control methods currently in use, as opposed to an eradication effort such as that instigated for the cotton boll weevil (*Anthonomus grandis*) in the Southern states. Ultimately, they are encouraged to conduct their own research to develop procedures that may help in the control efforts. This is a complex undertaking as numerous factors have to be taken into consideration so as not to adversely affect the environment or the bat populations. The bats rely heavily on migrating moths as part of their diet and, as such, are an integral part of the food web.

One particularly promising research project is being undertaken by two fifth-grade students at Kemp Elementary School in Bryan, Texas. They were especially intrigued by two aspects of the life cycle of the worm. First, that the female moth can locate by smell the corn silks on which she will lay eggs and second, that a natural predator of the moth in Texas is the Mexican free-tailed bat. Their question to be investigated: "Would moth egg-laying on silks be deterred or reduced if the silks were sprayed with a solution of water and bat guano?" If this proves to be the case, then the outcome of their research might be an environmentally friendly means of worm control with the side benefit of improved corn plant growth through the application of a natural fertilizer. But, regardless of the outcome, these students are addressing environmental concerns by seeking an alternative control method to the use of a pesticide. All of these efforts sparked a debate amongst fellow students in schools. They all learned about the pros and cons of each control method.

Figure 4.3. The program director watches as students at Kemp Elementary School in Bryan, Texas, carefully measure to create a spray solution of bat guano and water.

School Visits

An essential component of the program is to support the teachers and students during the academic year following teacher participation in the summer institute. The program director makes it a point to teach for at least one day in the class of each teacher and is always available via e-mail to answer questions. When in the vicinity, he returns to the USDA/ARS research lab at that time to keep the scientists in the loop. As an example, in the 2008-2009 school year he taught for teachers currently in the program but he also managed to teach for some teachers from previous years or ones who contacted him through the program web page. He interacted personally with 3,141 students ranging from 1st through 8th grade. The ethnic makeup of the students was 10.8% African American; 21.7% Caucasian; and 67.4% Hispanic all of whom were taught in 31 different schools. These visits allow the director to see the impact that the project has on students and supports the teachers in their efforts to promote hands-on, inquiry-based science.

The ethnicity of the students reached is partly a product of the location of the USDA/ARS labs that are selected for participation. The program was recently moved under the umbrella of the USDA/Hispanic Serving Institutions National Program (HSINP). For this reason, initial expansion has been into areas with large Hispanic populations.

Results: Student Research Presentation Day

The program culminates with a Student Research Presentation Day (SRPD) at the end of the year where student/group/class research work is presented. Parents are actively involved with their students during the year and invited to attend the SRPD along with the school principal and district science supervisors who signed off on the teacher's participation from the outset and offered necessary support during the year while facilitating the school visits by program staff. This end-of-year event is modeled after scientific meetings where posters and papers are presented. It is the test for "actual learning," gauged by the quality of the student work that is on display. Each participating teacher selects from four to five students who have taken the most interest in the program and produced either a power point or tri-fold poster representation of their research. They then present their work in front of an audience of their parents, peers, teachers, administrators, staff, and research scientists.

Over the past few years, it has been noticeable that there has been an equal representation of both boys and girls (Gurian and Ballew 2003; Hyde and Linn 2006) at the SRPD. The floor is open for questions during and after each student presentation and scientists always comment on the composure of the student presenters. Then there is a Scientists Question and Answer Panel where four research scientists volunteer to sit as a panel to field questions about their research from the audience. The type of questions asked usually covers a wide range of topics, from those directly related to the worm research to those relating to different local research programs. Before the floor is opened for questions, each scientist gives a brief summation of when and how he or she first became interested in science, how each eventually ended up working for the agency, and what sort of opportunities the job has afforded, from job satisfaction to world travel. This is a popular session and is followed by an equally popular catered meal.

In the afternoon, all participants take part in interactive tours of some of the more visual research projects currently being undertaken at the particular laboratory. In the past, one tour featured a fistulated cow at the USDA/ARS Southern Plains Agricultural Research Center at College Station, in Texas where scientists have installed a plastic device called a "cannula" into the side or stomach of an animal. This allows them access to the various organs and they can observe how fast the cow digests various foods and what chemical/biological processes the food undergoes. The students were invited to put their hand into the cannula to take a sample of the cow's stomach contents and to view the microorganisms contained in the sample under a microscope. On another tour, at the Cotton Ginning Research Unit in Las Cruces, New Mexico, scientists allowed each student to operate a mini-gin to gin several cotton bolls so that they could see the process and collect the seeds and separated cotton fibers. They were able to inspect a two-row cotton picker up close and see how it is being adapted for pepper harvesting, a process traditionally carried out by hand, which is both labor-intensive and expensive. If the process can be mechanized, it will allow American pepper growers to remain competitive with foreign imports. Tours like this increase teacher and student understanding that science and technology research is interwoven with many other complex issues.

The Promotion and Dissemination of Components of the Program

During this portion of the unit, the program director makes presentations at educational and scientific conferences. For example, in 2008–2009, the director introduced the project to teachers and undergraduate students in Arkansas, California, Colorado, Indiana, North Carolina, Oklahoma, and Texas. A genuine attempt is made to attend conferences aimed at underrepresented minorities but also at conferences that target teachers already heavily invested in life and agricultural sciences. It was especially rewarding to present at the Texas State Science Teacher Conference in Fort Worth to 120 teachers and to interact with the American Indian Science & Engineering Society (AISES) undergraduates in Anaheim, California, at the national conference and then again at the AISES Region IV conference in Tahlequah, Oklahoma. Other exciting venues include the Hispanic Engineering, Science, and Technology (HESTEC) week in The Rio Grande Valley; the national conference of the Society for the Advancement of Chicanos and Native Americans in Science (SACNAS); and the national conference of the Hispanic Association of Colleges and Universities (HACU).

The Multiplier Effect and Worm Shipment Data

The demand for insects reached a point where it was beyond the capacity of the College Station Area-wide Pest Management research unit. As a result, Wayne Ivie, Southern Plains Area Research Center (SPARC) Director and John Westbrook, the Area-wide Pest Management Research Unit (APMRU) Leader in College Station, hired a student as a Biological Science Aid with the primary responsibility of filling teacher caterpillar requests in a timely manner. This was a welcome and essential addition to the program and so far this year, 16,561 caterpillars have been shipped to be studied by 12,476 students from grades preK–12.

The number of worms that have been shipped has risen steadily each year since shipments began in 2005. The service is free, thanks to the generosity of SPARC. The eggs are supplied by the USDA/ARS, Crop Science Research Laboratory, in Mississippi, through the Genetics and Precision Agriculture and the Corn Host Plant Resistance Research Units, and from the Monsanto insect rearing facility in Tennessee. This ensures a continuous supply of eggs, again at no cost to the program. This is a case of genuine collaboration between partners in an educational outreach effort that values the teachers' efforts to engage and to create enthusiasm among their students for science.

Future Scientists Website

An interactive project website was developed at USDA/HSINP in Washington DC (*www.hsi.usda.gov/CornEarWorm/main.htm*). The site has a data-entering component—a real-time tracking system called PestWatch—where individual classes can enter worm growth data and compare their data with that from other schools in the same district or look further afield to data from schools within their state or from other states. In addition, there is a list of video links of, for example, time lapse video of corn growing and the Iowa Farmer Corn Cam.

Future Scientists Program Extension

The current Future Scientists Program targets students in grades 4–12 with an emphasis on the middle grades where research indicates students begin to lose interest in science. For example, the 2005 National Assessment of Educational Progress science results show steadily declining student achievement in science between grades 4 and 12. In that national assessment, 68% of fourth graders performed at or above the basic achievement level, compared with 59% in grade 8 and 54% in grade 12 (Nagel 2007).

In order to extend the outreach effort, a pilot program was undertaken in summer 2007 when two Texas A&M University undergraduate students were mentored by Dr. Mike Grusak from the USDA/ARS laboratory at the Children's Nutrition Research Center (CNRC) in Houston, Texas. Then, in the summer of 2008, nine undergraduate students in the Texas A&M College of Science were given the opportunity to work in the USDA/ARS/Southern Plains Area Research Center (SPARC) in College Station to learn about and to participate in the research process. The Undergraduate Research Internship (URI) is undertaken for credit rather than monetary compensation and comes under the umbrella of Biology 491: Independent Research, a research course supervised by various faculty members in the various Colleges (1–4 credit hours). In addition, eight credit hours may be used as a direct elective. Additional credit hours may be used as general electives. If credit is not required, students are welcome to participate for the research experience. This program was again offered in summer 2009, with seven scientists designing specific projects for students to work on at the SPARC facility.

What better way for aspiring research scientists to gain experience in their field of choice than to be mentored by experts in their fields with a passion for sharing their knowledge? These are all leaders in their fields and are willing to share their research expertise and to share their perspectives on matters of public discourse and debate surrounding their sometimes sensitive research projects.

Evidence of Successes

It was decided at the outset that the prime measure of success of the program would be the final student work that was presented at the end of the year. These are known as the Student Research Presentation Days. They are the culmination of a yearlong working relationship established between the scientists, teachers, and students, and as suggested by Drayton and Falk, succeed because of the initial "scaffolding" provided to the teachers (2006). The high standard of the presentations, delivered by confident and poised student presenters, were a revelation to the research scientists. The projects varied in the length of time spent on data collection; but all students were engaged for at least six weeks in the project (normal time span for the life cycle of the insects). These presentations represented the full range of variations of the "Essential Features of Classroom Inquiry" (NRC 2000, p. 29) and ranged from more teacher-directed and guided, through the teacher being responsive to student suggestions and input; to the ideal where the research activities were totally student-centered and driven by their questions, interests, and enthusiasm.

Teacher guidance was often more pronounced at the lower grades where projects ranged from basic observation of the life cycle and research to determine if the insect was either a moth or butterfly, to experiments to try to determine if the worms were cannibalistic (they are). These experiments were often conducted as class or group projects but teacher flexibility meant that individual students could pursue independent studies. The following examples, by grade level, of some previous research projects highlight the breadth and depth of study.

Grade 2

The teacher attended the summer institute at the USDA/ARS Jornada Experimental Range and Cotton Ginning Research Unit in Las Cruces, New Mexico, in 2005. She came to the program as a fifth-grade teacher but was reassigned by the school to teach second grade at the start of the 2005 school year. She adjusted her lessons accordingly as we agreed to support her in bringing the corn earworm research into her new situation. By the end of the year, her students had successfully grown the worms through the complete life cycle, created puppets of each of the stages, and put on a puppet show of the life cycle for the rest of the school. One particularly creative student made a large, illustrated comic book detailing the life cycle in a humorous vein; he presented this effort at the Student Research Presentation Day.

Grades 3–5

A third grader in Lubbock, Texas, was able to pursue an answer to his question, "Can I disguise the scent of the corn silks to deter egg-laying of the moth?" A fifth-grade girl In Las Cruces, New Mexico, experimented to learn, "Will a thinner skinned kernel be preferred by the worms over a thicker outer skin from a different type of corn?" A group of fifth-grade students near Beaumont, Texas, were studying beneficial Human Nutrition and turned their attention to the diet of the worms to explore if the type of diet impacted the worm's life cycle. They designed their own Food Energy Experiment (see this chapter's appendix) in the standard research design format.

Middle School

Two girls from the middle school in Junction, Texas, decided to pursue their own research question: "Can we successfully determine moth gender at the pupa stage?" The girls first presented their project at the Junction Middle School Science Competition where it took first place, and again at San Angelo, Texas, at the Regional Science and Engineering Competition where they took first place in the Life Science category for teams. They traveled to the State Science and Engineering competition in San Antonio, Texas, and were invited to participate in the Discovery Young Scientists Competition as well as garnering an Agricultural Award. While their research will not advance efforts to control the worms in the field, it achieved an equally important goal: intriguing two students enough to ask a question that interested them and then to pursue an answer.

High School

Basic Student Research

As an example of one research project, 10th-grade biology students at Valley Springs High School in Arkansas developed a project over an extended period and ordered worms over several months. Their work exemplified "open inquiry." They had several ongoing projects but the initial problem investigated was, "How long is the life cycle of the corn earworm? How long is the growth rate of larvae? Can we produce a second generation?" They observed the growth of the worms while recording data on a daily basis, but were unsuccessful in producing a second generation. They thought they had done everything correctly, so they researched factors that affect reproduction and consulted the lead corn earworm USDA/ARS Research Entomologist, Dr. Juan López,. They knew that during the day, their classroom was kept at a constant temperature (20–22°C) but found out, by chance, that the janitor liked to work in "arctic conditions." They now knew that the worms were subjected to cold temperatures and solved this problem by adding a warming pad at night to the worm containers. This allowed them to be successful in producing a second generation and from there they moved onto more complex investigations on the effect of temperature change on the length of the life cycle to see if the worms might be controlled by chilling the field crop in some manner.

Along the way, they also had practical experience of some of their worms entering diapause and learned about sexual dimorphism, first by their own research data and then by online research. Their teacher had driven several hours to be able to attend the summer institute at USDA/ARS Harry K. Dupree National Aquaculture Research Center in Stuttgart, Arkansas, in 2006. She has successfully studied the worm in each successive year with new students and has expanded student involvement with input from her students from previous years.

Award-Winning Student Research

Some students progress to county, state, and then national science competitions. For example, in 2008, a high school student from Alpena High School in Arkansas placed third in the nation at the Future Farmers of America (FFA) Agri-science Fair. Her project, in the biochemistry category with work on *Helicoverpa zea*, highlighted the current public discourse and debate over organically grown foods and those foods that are treated with harmful insecticides. She hypothesized that the common preservative, sodium benzoate, would have an adverse affect on the lives of corn earworms while not causing concern about human consumption of the corn. The experiment also tested different levels of the preservative, and each level's affects on the corn earworms. The data and observations suggested her hypothesis was correct. Therefore, farmers might be able to use the preservative as an insecticide to control insects and other pests on their crops instead of the regular insecticides. To continue the research, the student planned to track the growth of the adult corn earworms that develop and make comparisons to see if the preservative makes a difference in their development and in their growth rates, compared to the worms in the control group.

Published Research

The ultimate test of learning within the scientific community was achieved by a Bryan, Texas, high school senior who was senior-author with Texas A&M University and USDA/ARS scientists on a published paper: Effects of *Bacillus thuringiensis Kurstaki* and Sodium Bicarbonate in Coleopteran and Lepidopteran Larval Diets (Skrivanek et al. 2006). This is the abstract from *Southwest Entomologist*:

We tested the hypothesis that *Bacillus thuringiensis* subspecies *kurstaki* (*Btk*) would adversely affect coleopteran larvae if the mid-gut pH is adjusted to the alkaline pH of lepidopteran larvae, which are susceptible to *Btk*. Corn earworm or Bollworm, *Helicoverpa zea* (Boddie) (Lepidoptera: Noctuidae), and mealworm, *Tenebrio molitor* L. (Coleoptera: Tenebrionidae), larvae in TRT 1 were fed Instant Soybean-Wheat Germ Insect Diet (SWD); TRT 2, SWD + sodium bicarbonate (to alter the mid-gut pH); TRT 3 SWD + *Btk*; TRT 4, SWD + sodium bicarbonate + *Btk*. Mealworms in TRT 1, 2, and 3 all lived and had a comparable weight gain. Mealworms in TRT 4 had some mortality and no weight gain. Bollworms in TRT 1 all lived and gained weight. Bollworms in TRT 2 had low mortality and gained weight. Bollworms in TRT 3 and 4 had high mortality and no weight gain. TRT 2 and 4 containing sodium bicarbonate significantly increased mid-gut pH in bollworm and mealworm, changing the latter from acidic to alkaline. Our results supported the hypothesis that coleopteran larvae with an alkaline mid-gut would be adversely affected when fed *Btk* and also suggest that increasing the alkalinity of bollworm mid-gut increases susceptibility to *Btk*.

This is indeed an example of a student working on a local issue that had personal relevance to her and had both local importance and the potential to have a national impact.

Program Publicity

Dr. Wilson gained positive exposure for the program by including an educational component in a marathon run in Austin, Texas, on February 15, 2009. He took a corn earworm with him and asked all 1,250 Bryan ISD fifth-grade students to decide if his running "buddy" was a male or female. After the marathon, the data of their guesses were compiled and feedback provided to students after the worm entered its pupa stage when gender could be determined. This story is available at *www.science.tamu.edu/articles/639*.

Conclusion

Without exception, there was great cooperation and support for the summer institutes and the yearlong follow-up by all of the stakeholders, culminating in the Student Research Presentation Days. The science learning community being developed is built on genuine collaboration. The USDA/ARS research scientists who were asked to share their research with teachers gave generously of their time and were more than willing to share their expertise. The USDA/ARS administration and CMSE staff both worked to make each institute a success. This depth of implementation support, the thoroughness of the original inservice in preparing the teachers (Kyle, Bonnstetter, and Gadsden 1988) and incorporating them into the culture of the USDA/ARS research laboratories during the summer, and finally, the exposure to inquiry were all essential to success. The program allowed the teachers to "deepen their understanding of the nature of science as a creative, knowledge-making process" (Drayton and Falk 2006, p. 253). Importantly, the students rose to the challenge and often exceeded expectations in conducting their research and, just this year, a student who had attended the Student Research Presentation Day at the USDA/ARS Southern Plains Agricultural Research Center at College Station as a Fifth Grader in 2004, has applied to work there at a summer job. He is now a junior in high school.

The following quotation sums up the promise of the Future Scientists Program: "Observing and conducting research on the corn earworm is a delicate process but, if you are careful, you can have a lot of fun. If you are observant enough, you can learn things that you never thought were possible. Thank you for inviting us to participate." (Snook ISD, Texas, sixth graders.)

Through their involvement in the Future Scientists Program, students (K–12) are provided a model that enables them to begin to "Engage intelligently in public discourse and debate about matters of scientific and technological concern" by examining the question "Who ate our corn?"

Acknowledgments

This work was funded by the USDA/ HSINP, USDA/ARS/SPA, and Texas A&M University Center for Mathematics and Science Education (CMSE). The opinions expressed herein are those of the authors and not necessarily those of the funding agencies. In addition, we would like to acknowledge Chris Parker and Parker Knutson for their invaluable work is preparing shipments of worms for schools.

References

Abruscato, J. 2000. *Teaching children science: Discovery methods for the elementary grades and middle grades.* Needham Heights, MA: Allyn and Bacon.

Alberts, B. 2008. Editorial: Considering science education. *Science* 21 (319): 1589.

Beck, M., E. Morgan, S. Strand, and T. Woolsey. 2006. Mentoring: Volunteers bring passion to science outreach. *Science* 24 (314): 1246–1247.

Drayton, B., and J. Falk. 2006. Dimensions that shape teacher-scientist collaborations for teacher enhancement. *Science Education* 90 (4): 734–761.

Gurian, M., and A. Ballew. 2003. *The boys and girls learn differently action guide for teachers.* San Francisco, CA: Jossey-Bass.

Hyde, J. S., and M. C. Linn. 2006. Gender similarities in mathematics and science. *Science* 27 (314): 599–600.

Kyle, W. C., R. J. Bonnstetter, and T. Gadsden. 1988. An implementation study: an analysis of elementary students' and teachers' attitudes towards science in process-approach vs. traditional science classes. *Journal of Research in Science Teaching* 25 (2): 103–120.

Lederman, L., and S. Malcom. 2009. Editorial: The next campaign. *Science* 6 (323): 1265.

Ledley, T., N. Haddad, J. Lockwood, and D. Brooks. 2003. Developing meaningful student-teacher-scientist partnerships. *Journal of Geoscience Education* 51 (1): 91–95.

Lempinen, E. 2006. Education: Senior scientists and engineers bring experience to class. *Science* 30 (312): 1891–1892.

Nagel, D. 2007. STEM Education: Prepping students for science. *The Journal. www.thejournal.com/articles/21366*

National Academies Committee on Science, Engineering, and Public Policy (COSEPUP). 2005. Executive summary. *Rising above the gathering storm: Energizing and employing America for a brighter economic future.* Washington, DC: National Academies Press.

National Research Council (NRC). 1996. *National science education standards.* Washington, DC: National Academies Press.

National Research Council (NRC). 2000. *Inquiry and the national science education standards.* Washington, DC: National Academies Press.

National Science Resources Center (NSRC), National Academy of Sciences (NAS) and Smithsonian Institute. 1997. *Science for all children: A guide to improving elementary science education in your school district.* Washington, DC: National Academies Press.

Omenn, G. 2006. Presidential address: Grand challenges and great opportunities in science, technology, and public policy. *Science* 15 (314): 1696–1704.

Pfund, C., C. M. Pribbenow, J. Branshaw, S. M. Lauffer, and J. Handelsman. 2006. The merits of training mentors. *Science* 27 (311): 473–474.

Russell, S. H., M. P. Hancock, and J. McCullough. 2007. The pipeline: Benefits of undergraduate research experiences. *Science* 27 (316): 548–549.

Scott, T., C. Schroeder, H. Tolson, and A. Bentz. 2008. *Effective K–12 science instruction: Elements of research-based science education.* Pamphlet for ISDs.

Skrivanek, S., B. Ripple, J. Lopez, and M. Harris. 2006. Effects of Bacillus thuringiensis Kurstaki and sodium bicarbonate in coleoptera and lepidopteran larval diets. *Southwest Entomologist* 31 (1): 55–58.

Texley, J., and A. Wild, eds. 1996. *NSTA pathways to the science standards. High school edition.* Washington, DC: National Academies Press.

Wilson, H. C. 2001. A multiple case study: The perceptions and experiences of four research scientist-science teacher teams in a scientific work experience program for teachers (SWEPT). Unpublished doctoral dissertation. College Station, TX: Texas A&M University.

Appendix

Question:
Will changing the worms' diet affect the number of worms that will be able to pupate and become adults?

Hypothesis:
Assuming that worms need a balanced diet like humans to be healthy, then the low-energy-diet worms may not pupate.

Materials:
24 worms in individual cups with regular diet
Alternative food of fresh lettuce; and slices of tomato (to be changed daily) with tissue paper to be placed under tomato to soak up liquid.
(The students understood that the worm feeds on tomato as the tomato fruit worm.)

Procedures
Allow the worms to stay on original diet until they reach the second in-star stage.

1. Gather 24 worms about the same size for the experimental groups.

2. Put 8 worms on a low-energy diet of only lettuce. This is group 1.

3. Put 8 worms on a medium-energy diet of ripe tomatoes. This is group 2.

4. Leave 8 worms on the high-energy diet that was sent with them. This is group 3.

Data:

Group Number	1	2	3
Number pupated			
Days to pupate			

Conclusion:
The students were able to present their findings in May at the Student Research Presentation Day. Both sets of worms on high- and medium-energy diets pupated successfully. However, the worms avoided the lettuce so the students decided that they would have to redesign the experiment using newly developing "cotton squares" as the food source, as the worm can also be called the cotton boll worm.

This provided a valuable lesson, because sometimes research does not go as planned and it has to be rethought and redesigned.
Other research projects are available at: *www.hsi.usda.gov/CornEarWorm/main.htm*

Applications of Biology as Part of a Preservice Program for Science Teachers

Hakan Akcay
Marmara University

Setting

The University of Iowa has been studying and experimenting with varying patterns for a preparatory program for new science teachers since gaining the U.S. Department of Education funding in 1993. There were three Salish grants between 1993 and 1998. The Salish efforts continue today with funding from the National Science Foundation for IMPPACT (Investigating the Meaningfulness of Preservice Programs Across the Continuum of Teaching) with Syracuse University as the lead institution. One of the innovations at Iowa was the introduction of applications courses in each of the major science disciplines—first tried as part of the Iowa Chautauqua Program for inservice teachers in 1982.

As a doctoral student at the University of Iowa, I was privileged to assist with the Applications of Biology course—first as an assistant and later as the primary instructor for two years. Although now a professor at Marmara University in Turkey, I introduced what I learned at Iowa about applications courses and have been approved to establish such courses at my university. The first attempt has been for biology—conceptualized and approved for introduction in 2010. We hope to collect similar data to that reported in this chapter and examine it for future reports.

The Salish research ended with a series of recommendations thought to provide new directions for preparation of reform-minded teachers. The program features recommended include:

1. a strong science component consisting of a major science area (30 + semester hours (s. h.)) and a supplementary science area (at least 15 s. h.);

2. specific work in the history, philosophy, and sociology of science (6 s. h.);

3. an introduction to the design world (technology) and its ties to the natural world (6 s. h.);

4. at least two experiences (one semester each) with applications of major science disciplines, including work with current issues (guiding questions, possible explanations, selection of "best fit" solutions, corrective actions);

5. a science and a technology research experience;

6. a three-semester methods sequence and each with a practicum experience in schools over a three semester period providing experience in elementary schools, middle schools, and high schools;

7. a full-time student teaching for one full semester;

8. university general education requirements for Bachelor Degrees (about 30 s. h.); and

9. general education (philosophy of education, educational psychology, special education, diversity, action research) (10 s. h.).

Over 20 universities were involved in trying some of these recommended changes after the initial Salish project. All were introduced and tried by varying faculty members at Iowa over the decades that followed.

At the University of Iowa there are four applications courses in each of the major disciplines (biology, chemistry, physics, and Earth science). These are considered more important than enlarging the quantity of either science content experiences (via teacher lectures, textbooks, and demonstrations) or with desired pedagogical knowledge. Instead the courses were created to be sure teachers have the ability to use the typical concepts (content and/or pedagogical) in new contexts. The applications of science sequence consist of courses that focus on the societal and educational aspects of each discipline. Students use the information gained from the teacher education preparation program in the sciences to plan ways to address societal issues in our world. They also learn to use information to identify issues, collect data, develop arguments in support of a position, and take appropriate actions in order to resolve the issue. This is a study concerning the Applications of Biology course on the university campus for preservice teachers—both the course I taught at Iowa and the one I am preparing for this academic year at Marmara University.

Studying the Impact of the Applications Approach

The effectiveness of the Applications of Biology program on participating students was studied during two semesters. There were 41 (13 males and 28 females; 19 elementary science majors and 22 secondary science majors) preservice science teachers enrolled in Societal & Educational Applications of Biological Concepts. An interested graduate student colleague assisted with data collection and analyses. Data collection for the sample in Turkey will be collected throughout the 2010–2011 academic year. It is my intention to compare results of such applications in other countries.

Unique Features of the Course

The course is a one-semester, four-credit course. The instruction included one three-hour weekly session for 14 weeks. Most work on student projects was accomplished outside of the scheduled class time. The nature of the course requirements and experiences indicated in the course syllabi involve all students in identifying specific research questions where their knowledge of biology could be used. Table 5.1 is a listing of the facets of the course and the requirements for each student to complete. The changes in students as well as their reactions to the course indicate that the applications courses contribute significantly to the education of new science teachers. The goal of the courses is to assist in relating traditional science content to the recommended (and research-based) methodologies.

Table 5.1. Organizers for Class Sessions for the Applications of Biology Course

1.	Individual projects for each student illustrating the use of biology study for resolving personal and/or societal issues.
2.	Group projects (3–4 members) involving a local and current problem where biology study can be helpful in resolving it.
3.	Preparation of a paper to evaluate a Science-Technology-Society (STS) approach to teaching as used by hometown or local teachers.
4.	Critique of ten reports from popular press in terms of problems, resolutions, and potential actions.
5.	Analysis of a product advertisement where biology knowledge would be useful in creating (or refuting) the claims.
6.	Preparation of a literature review of program results and corrective actions with the use of STS as a teaching approach (K-16).
7.	Analysis of an individual project completed by another student in class.
8.	Elaboration of individual and group actions taken by another student and/or those individuals in another group.
9.	A report on what each student identifies as the "most important" actions taken by another student in the class regarding an individual or group project.

During the course, I encouraged a learning environment where students have an opportunity to study the interactions among science, technology, and society from a variety of contexts. They also explored how these systems affect each other and are affected by humans in general. Students were encouraged to learn how to make informed decisions about science and technology issues in a social context. This is in line with goals 2 and 3 of the National Science Education Standards: Goal 2) Use appropriate scientific processes and principles in making personal decisions and Goal 3) Engage intelligently in public discourse and debate about matters of scientific and technological concern (NRC 1996, p. 13).

Students in the applications of biology course were involved with various learning experiences, including role-playing, debates, library searches, brainstorming, problem solving, reports of individual and group projects, case studies, general class discussions, and presentations at various community organizations. The problems and issues studied were all identified independently by the students.

The course is focused on issues and questions raised by reports identified by students from television programs, articles in national and local newspapers, and writings in popular magazines. The first objective of the class was to formulate the questions that these reports suggested. The next requirement was to apply biological information and inquiry techniques to explore their questions. The issues identified, the questions formulated, and the answers proposed provided topics for class discussions and for individual and group projects. Personal actions were expected after the analysis of the issues. The issues and questions provided the vehicle for accruing much more than just typical science content; they provided motivation and opportunities for students to interact with their peers, teachers, school and community leaders, and experts, as the students searched for more information, considered alternative solutions, and actually applied these experiences to deal with a variety of real-world issues.

The major focus for the course is the identification of a biologically relevant local issue or problem. It must be a problem that is personally relevant and interesting. This local problem often comes from city council members, boards of supervisors, boards of education, or other community agencies. Each student analyzes the problem using information gained from talking with public officials, biologists, professionals (many with college backgrounds in biology), and members of the general public. Wherever possible each student is encouraged to get directly involved with the issue by taking stands, entering into debates, writing editorials, or collecting detailed information for those principally involved in the controversy. The emphasis, of course, is upon the *use* of biological information and scientific procedures. Student projects are to be designed to provide a direct experience of using biological concepts and scientific procedures in local communities and in actual decision-making situations.

I expect each student to prepare a term paper that elaborates on a position taken, recording the actions performed and evidence assembled. A summary analysis is needed that focuses on the results and significance of the activity. It is expected that each student will complete a literature search, have personal interactions with "experts," and take a survey of public opinion where differences are investigated—with age, education, socioeconomic background, and other factors noted, compared, and analyzed.

Group Projects

Another focus for the class is the identification of class projects where three to four students work cooperatively to accomplish goals identified by the class during initial class discussions. This effort is designed to illustrate cooperative learning, division of labor, use of various individual strengths in the group members, and group dynamics. It often includes activities associated with local agencies and citizens. This often means involvement with the legislature or a legislative committee in Johnson County or the City of Iowa City. Significant class time is used for presentations and discussions by officials, involving all class members with the issues explored by the small groups.

Since the class projects are group efforts, individual student evaluations are completed by all other students in each group. Three questions are asked regarding each project: What did each of the other students do? What were relative successes? How did the quantity and quality of involvement compare with others? Each member of each project group was also asked to evaluate the efforts of all other group members in the particular group. Table 5.2 is a list of examples of topics chosen by students for their individual and group projects over the two semesters.

Course Features

Each student was asked to critique 10 reports from the popular press (each 3–4 printed text pages) during the duration of the course. At least one of these was to come from *World Watch Magazine*, the journal of the World Watch Institute, which focuses on reports on the environment

Table 5.2. Titles of Student Individual and Group Projects Over Two Semesters[1]

Individual Projects	Group Projects
Agent Orange	Proposed National Environment/Energy
Menomune's Debate	Center
Illegal Performance Enhancers	Radon in the City
Hepatitis Vaccine	Date Rape Drugs
Parkinson's Disease	Global Warming
Entrapment Neuropathy	Alternative Energy Sources
Chronic Wasting Disease	Genetically Modified Foods
Fast Foods	Obesity in Young Children
Racial Discrimination	Energy Drinks
Eczema	Diabetes
Down's Syndrome	Pollution of City Water
Saving a life: MPS & BMT	Recycling
Grief & Despair vs. Depression	Cholesterol
Atkins Diet	Nutrition
Scoliosis	Cloning
Depression and Counseling Services	
Stopping people from smoking?	
Thyroid Disease	
CPR & AEDs	
Autism	
Organ Donation	
Adult Macular Degeneration	

[1]Readers should be reminded that the lists are abbreviated versions of titles from individual and group projects proposed by students—not teachers—and not from "suggested project booklets." Instead, all individual projects were selected because of personal ties or problems that were identified by students. Likewise, group projects had student ties and were of special interest. They also were to be locally based, and of current interest, also seen as problems by community leaders. In one sense all were tied to personal concerns for each student. Often proposed plans, actions, and resolutions resulted in documents larger than this chapter manuscript for each student. All class sessions were headed by students seeking assistance and involvement from others in the class to work for "their" projects or to get help from other students relative to their own problems (Akcay and Yager, Forthcoming).

from some of the world's leading environmental proponents. The final paper is an attempt to synthesize the news reports in terms of problem, resolution, and potential actions. The individual articles were to be referenced with specific quotes used in the analysis. The articles read were used to supplement the work on group projects and the individual project. Information in the final report was to include information relevant to the following questions:

1. Why do you consider the problem or issue identified as important?

2. What are some biological concepts, laws, and principles related to your problem and issues?

3. How do you propose to resolve the problem and issues?

4. What basic biological concepts, laws, and principles are used in your efforts for resolving the problem and issues?

5. What difficulties did you encounter when you were taking actions for resolving the problem or issue? What help from others did you arrange?

6. What specific steps would you take as a next action for resolving the problem or issue?

Another facet of the course was to look at education (any academic level) to determine what techniques were being employed to help learners (students) apply biological information (gained from the college work in biology). Some reviews of science/technology/society (STS) programs in elementary schools, secondary schools, and colleges were encouraged. The STS programs were seen as promoting the teaching and learning science and technology in the context of human experience. This means focusing upon current issues and attempting to resolve them provides the best way of preparing students for current and future citizenship roles. This means identifying local, regional, national, and international problems with students; planning for individual and group activities that address them; and moving to actions designed to resolve the issues investigated. The emphasis in the course is on responsible decision making in the real world of the student where science and technology are components. The approach provides a means for achieving scientific and technological literacy for all. Each student is expected to locate one exemplary program to be analyzed carefully in terms of course content and instructional strategies employed.

Assessing Successes

Class sessions at midterm and toward the end of the semester were devoted to student presentations of individual and group projects. Class members were expected to enter into discussions concerning all projects, offering suggestions, criticisms, and evaluations. Whenever possible these presentations were held at sites designed to inform and affect other people—those not enrolled in the class. The presentations often involved citizens, youth groups, business roundtables, government agencies, students in K–12 environments. Half of the class was assigned to complete the group reports at midterm and the other half at the end of the semester.

I planned two examinations. The first consisted of a written article that was selected by me. Students were asked to analyze it individually. Students were asked to plan actions and to illustrate their abilities to use biological concepts in specific situations. The second and final examination consisted of a critical review of one of the individual projects completed by one of the other students in the class.

Instruments and Data Collection

I used multiple data sources to investigate the preservice science teachers' beliefs about the applications approach, including (1) written reflections by students; (2) individual and group project reports; (3) analysis of classes where Applications approaches were used at other academic levels;

(4) analysis of two other final reports completed by other students; and (5) results from administering a constructivist teaching questionnaire.

Qualitative Data Collection

The first data collection techniques were written reflections by each student. Each week, at the end of the class session I asked preservice science teachers to respond anonymously to open-ended questions that were related to the class topics and discussions. A second data source included evaluations of student individual and group projects. The individual and group projects were designed to provide a direct experience of using biological concepts and scientific procedures in local communities and in actual decision-making situations. Each student was expected to prepare a term paper for his or her individual project that elaborated on a position taken, a record of the actions performed, and the evidence assembled to indicate level of success. A summary analysis was reported, which focused on the results and significance of the activity.

The changes in students as well as their reactions to the applications of biology course indicate what applications courses contribute to the education of teachers. The goal of the course is to assist in relating traditional science content to recommended (and research-based) methodologies. Other data sources were evaluations and challenges of all students regarding their individual and group efforts. Students did not merely listen and observe; they collected the practices and outcomes of the work of other students enrolled. Often they were asked to suggest other interpretations and uses for the products, including the appropriateness for elementary school, secondary school, and college. Each student was expected to locate one exemplary program and to analyze it carefully in terms of course content and instructional strategies employed.

The last qualitative data source was the final reports of the students. Each student was expected to review one of the other individual projects that were completed by one of the other students in the class. The paper should clearly identify the following:

1. What was the issue? How significant was it?

2. What source of information did the student use?

3. What biology that students already knew proved useful to them in analyzing the project?

4. What new biological knowledge did they learn? How? What was the source of this information?

5. How do students evaluate the public poll that was used? Was it effective? Are there any problems? What are some alternatives that might have been used in this area?

6. What actions were taken or planned? How do students evaluate this action in terms of good citizenship? What alternative actions did students suggest?

7. What personal actions did students undertake after analyzing the individual projects in more depth?

8. What implications do students see for STS-oriented studies? How can more students gain more direct experiences from such courses?

Qualitative Data Analysis

The preservice science teacher conceptions of the applications approach were organized in terms of written responses to open-ended questions (written reflections as well as individual and group project reports), analysis of classes where the applications approach was used at other academic levels, and analysis of two other final reports completed by other students. The interpretative-descriptive approach described by Strauss and Corbin (1990) and the methods involved with constant comparative analysis as described by Glaser and Strauss (1967) were used. They were all used to analyze the qualitative data.

Quantitative Data Source

Quantitative data were collected by administrating the Constructivist Teaching Practice Scale (CTPS). Yager (1991) developed the CTPS earlier to enable teachers or students to reflect on their own experiences concerning use of constructivist teacher practices. It is a self-check instrument for determining to what extent a teacher or a course allows/encourages students to construct their own meanings. It can also be used to determine the degree to which a given instructor is utilizing an applications approach in teaching science courses. The extent to which a teacher allows students to construct their own meanings varies among teachers, individual students, and particular class settings. The CTPS includes fifteen statements characterizing teaching and learning from an "applications viewpoint" (see Appendix A). Students rated these statements in terms of their perceptions of their experiences in the class. The CTPS was administered to the students at the end of the semester along with an instructor evaluation form. Data gathered from the CTPS administration were analyzed for each of the items, using frequency distribution to characterize trends in the respondents' perceptions of the class.

Evidence of Impact From the Applications Courses

The Applications of Biology course does a great job of actively involving students in seeking information that can be applied to solve real-life problems beyond the class period, the classroom, and the whole university. It allows students to investigate, analyze, and apply concepts and processes to real situations. A good program has built-in opportunities for the students to extend beyond the classroom and to their lives beyond the campus. This new focus on learning helps to provide the basis for empowering students as future citizens to realize that they have the power to make changes and take responsibilities for doing so. The applications approach also empowers students to develop skills that allow them to become active, responsible citizens by responding to issues that impact their lives. Some typical examples of student comments concerning the new applications approach include:

> It is very open to your interests and connecting them to the community.

> It gives the individual more control over what they learn and provides them the opportunity to research and take action over societal issues.

> It was a very good experience. I learned a lot about many important topics in our society.

I really enjoyed this class. Class involvement was great and I learned a lot based on my opinions. I found additional ways to understand biology and make it fun.

Preservice science teachers indicate that the applications of biology class helped them become involved with biological concepts that relate to local, national, and global issues. Examples include:

This course helped me while applying my knowledge of biology to outside, real-world issues. I also built new knowledge to add to the old.

In general, I think that issue oriented studies are a great way to get interested in the subjects that they are studying. I know I learn better when I am interested in what I am studying. The easiest way to make sure that kids are interested in what they are studying is to allow them to choose what they study. I know that the standards talk about several things, but one of the points is that it is hands-on and student-centered. I don't think you could get more student-centered, or hands-on than what happened in this course. Also by studying different personal issues, students can become aware of how science, more specifically, biology, relates to their own personal lives. I know that many times what is being taught seems like a whole lot of useless information that could never really be applicable to the real world. By starting off with the issue, there is no doubt that it really relates to the real world. Overall, I think that learning science by studying present issues or personally relevant issues is a great way to learn. It is an interesting way to learn and it would help make science classes interesting instead of boring. No more teachers lecturing at the students, but the students in a way become their own teacher.

The use of an applications approach in biology is a great way to get students motivated to learn. Students are able to relate to the social issues since they are directly related to their own lives along with those in their community. If students are able to choose their own issues, it is of greater interest to them. Students having opportunities to participate actively in their learning is a great motivational technique. When students get involved through researching, interviewing, working with the community, gathering data, and carrying out action plans, they are more motivated to learn because they have an active role and a precise reason. Direct instruction where students listen to a teacher lecture hardly provides anything about which the students can get excited. My preservice science teachers indicate that the applications course is student-centered and motivational for learning science. Specific examples from my students concerning their experience in my applications course include:

This was a very student-centered course. It was definitely my favorite class of the semester.

Issue-oriented research encourages very positive student driven lessons.

The students pick the topics; the topics are personally relevant to the students (and society). Students evaluate each other. I'd say that I enjoyed the projects, but I had to spend a lot of added time for them.

Social issues in biology are a great way to motivate learners. It completely addresses the relevance of the subject in everyday life. I feel that students, at any level, can be motivated by anything that they can relate to on a personal level.

By bringing in social issues in biology classrooms it allows the teacher something to work with—motivation. The motivation that students receive from seeing how biology is tied to their lives locally is something that can have great power in the classroom. Once students are engaged and want to learn with a topic that is relevant, it is much easier for teachers to then incorporate the concepts and processes related to the issue and getting other experts than "the" teachers involved.

The applications approach promotes social and communication skills by creating a classroom environment that emphasizes collaboration and exchange of ideas with others. Students must learn how to articulate their ideas clearly as well as to collaborate on tasks effectively by sharing in group activities. Students must exchange their ideas and must learn to interact with others. Moreover, they evaluate their contributions in a socially acceptable manner. This is essential to success in the real world, since they will always be exposed to a variety of experiences in which they will have to cooperate and navigate among the ideas of others. Some statements indicating this from my students include:

Honestly, I've never enjoyed doing projects as much as I have in this class. Every project was extremely interesting. I felt like getting out of other classes early just so I could work on my projects.

Work on how to inspire self-learning instead of just displaying and repeating information.

I would say I had a very positive experience in this class with surveying the public and learned a lot about how biology affects our lives.

Preservice science teachers' reflections on written responses and the assignments are not only positive but they also include a new appreciation for both how we learn science and how it should be taught. The following responses reflect this attitude:

Not only did I learn more about biology in specific cases (through individual/group projects) but I was also able to learn a new way to teach, learn, and think. I was first afraid of not having specific guidelines for everything but now I find that I would prefer it this way.

In comparison to how I was taught science previously, the applications method is completely different. I was taught all science by way of textbook and traditional methods, especially biology. Biology was always about memorization.

An interdisciplinary approach prepares and promotes students to become responsible citizens, ready to deal with an increasingly technological world. The applications approach leads to improved learning in the classroom and provides an example of interdisciplinary teaching. Preservice science teachers indicate that the applications approach is an example of interdisciplinary learning. An example from one of my students states:

The applications approach is a very complex subject and there are numerous questions that can be asked about it. Although it may be hard to believe, it does a great job of bringing both interdisciplinary learning and non-interdisciplinary learning to the classroom, and I can't wait to use it in my classroom one day.

According to the preservice science teachers' responses, there are several benefits of constructivism as a theory of learning. They indicated that they learn more, and enjoy learning more, when they are actively involved, rather than being passive listeners. Education works best when it concentrates on thinking and understanding, rather than on rote memorization. STS/Constructivist ideas concentrate on learning how to think and understand. Students also use the science information to solve real problem because constructivist learning is transferable by nature. STS/constructivism provides free learning environments for students. In constructivist classrooms, students create organizing principles that they can take with them to other learning settings. Some examples indicating these observations directly from my students:

I wasn't ready for the freedom. I came in expecting to do more with biology in the classroom.

I was surprised soon after I registered for this course because I had a completely different picture in my mind as to what it would be like.

It took a while to adopt to the freedom of the STS methodology and philosophy, but after that it was fun.

Preservice science teachers also indicated that society as a whole often does not pay attention to the issues that they considered. It seems that some students often are not comfortable in taking action in the broader society. An example from a student is:

In my classroom I would prefer to implement a constructivist approach mainly. I would like the Applications approach but I feel like I would only use it occasionally simply because I am not totally sold on the taking actions more broadly in society. I really support group learning and presenting materials to peers; for that reason cooperative learning and casual oral presentations will be frequent. I admire the idea of taking actions in society but in my experience society really does not care to listen or take much action based interest on what I have presented to them.

Scientific Quantitative Data Analysis

Data gathered from the administration of Constructivist Teaching Practice Scale (CTPS) were analyzed for each of the items, using frequency distribution to characterize trends in the respondents' perceptions of the class. Tables 5.3 and 5.4, page 80, indicate the numbers and percentages of students who responded in varying ways to each statement. These tables indicate that students found the Applications of Biology course to be student-centered and exemplified an authentic learning environment. It provided many learning opportunities for students to learn science. The results also indicate that the course achieved the goals that were elaborated prior to the course and evaluations at the end of the semester.

Conclusion

The Applications of Biology course provides students with the opportunity to apply scientific information in making decisions about real-world issues. Within the context of the course, students were introduced to an applications approach, with as many class examples as there were students. With such examples, students discussed instructional strategies needed to implement such a new

Table 5.3. Preservice Science Teachers' Perceptions of Varying Experiences in the Applications of Biology Class: Part A

	Student		Student-Teacher		Teacher	
	Number of Preservice Science Teachers	%	Number of Preservice Science Teachers	%	Number of Preservice Science Teachers	%
Identifies the Issues/Topic	24	63.2	9	23.7	5	13.2
Asks the Questions	21	55.3	14	36.8	3	7.9
Identifies Written and Human Resources	21	55.3	12	31.6	5	13.2
Locates Written Resources	24	63.2	13	34.2	1	2.6
Contacts Needed Human Resources	23	60.5	8	21.1	7	18.4
Plans Investigations and Other Activities	21	55.3	11	28.9	6	15.8

Table 5.4. Preservice Science Teachers' Perceptions of Varying Experiences in the Applications of Biology Class: Part B

	NO		N/A		YES	
	Nbr	%	Nbr	%	Nbr	%
Issue Is Seen As Relevant	0	0	4	10.5	34	89.5
Teachers Use Varied Evaluation Techniques	2	5.3	8	21.1	28	73.7
Students Use Varied Evaluation Techniques	6	15.8	7	18.4	25	65.8
Students Practice Self-Evaluation	3	7.9	4		31	81.6
Science Concepts Are Applied to New Situations	0	0	6	15.8	32	84.3
Science Skills Are Applied to New Situations	0	0	3	7.9	35	92.1
Students Take Action(s) as a Result of Their Work	1	2.6	4	10.5	33	86.9
Science Concepts and Principles Are Sought Out and Used By the Students	1	2.6	2	5.3	35	92.1
Evidence Available That Shows Science Learning Has Impact Out of School	0	0	3	7.9	35	92.1

instructional approach effectively in the teaching and learning of science. Issues provide a vehicle for acquiring more than just traditional science content. The issues are motivating and provoking as well, providing students opportunities to interact with their peers, other faculty members in the schools, and community experts. The applications course promotes scientific literacy.

The applications approach helps students to see science teaching and learning in a new way, one that encompasses all aspects of learning. When science is addressed in this way, preservice science teachers realize that they do not have to be the "knowers of all things" but rather, they become facilitators who assist students in investigating possible answers to their own questions. Science becomes not just a subject that is taught at a given period every week. Instead, it is seen as an all important subject that can be incorporated into every aspect of the curriculum. Teacher education programs should provide learning environments for preservice teachers that improve their understanding of the interactions of science, technology, and society.

References

Akcay, H., and R. E. Yager. Forthcoming. Accomplishing the visions for teacher education programs advocated in the national science education standards. *Journal of Science Teacher Education*.

Glaser, B. G., and A. L. Strauss. 1967. *The discovery of grounded theory*. Chicago, IL: Aldine.

National Research Council (NRC). 1996. *National science education standards*. Washington, DC: National Academies Press.

Strauss, A., and J. Corbin. 1990. *Basics of qualitative research: Grounded theory procedures and techniques*. Newbury Park, CA: Sage.

Yager, R. E. 1991. The constructivist learning model: Towards real reform in science education. *The Science Teacher* 58 (6): 52–57.

Appendix A

The Constructivist Teaching Practice Scale (CTPS)

Complete this evaluation form. Please mark each line with an X at the point on the continuum that best indicates your experiences in this course.

Teacher	IDENTIFIES THE ISSUES	Student
No	ISSUE IS SEEN AS RELEVANT	Yes
Teacher	ASKS THE QUESTIONS	Student
Teacher	IDENTIFIES WRITTEN AND HUMAN RESOURCES	Student
Teacher	LOCATES WRITTEN RESOURCES	Student
Teacher	CONTACTS NEEDED HUMAN RESOURCES	Student
Teacher	PLANS INVESTIGATIONS AND OTHER ACTIVITIES	Student
No	TEACHERS USE VARIED EVALUATION TECNIQUES	Yes
No	STUDENTS USE VARIED EVALUATION TECHNIQUES	Yes
No	STUDENTS PRACTICE SELF-EVALUATION	Yes
No	SCIENCE CONCEPTS ARE APPLIED TO NEW SITUATIONS	Yes
No	STUDENTS TAKE ACTION(S) AS A RESULT OF THEIR WORK	Yes
No	SCIENCE CONCEPTS AND PRINCIPLES ARE SOUGHT OUT AND USED BY THE STUDENTS	Yes
No	EVIDENCE AVAILABLE THAT SHOWS SCIENCE LEARNING HAS IMPACT OUT OF SCHOOL	Yes

(Yager 1991)

Linking Science, Technology, and Society by Examining the Impact of Nanotechnology on a Local Community

Joseph Muskin
University of Illinois
Next Generation School

Janet Wattnem
Mahomet Seymour School District

Barbara Hug
University of Illinois

Setting

n classroom conversation, a group of students talk about a new washing machine that uses nanotechnology to clean and sterilize clothes. The conversation follows:

S1: So this laundry machine, it puts in silver nanoparticles to like, clean the clothing that it washes?

S2: Yeah, I think so. I think I have the washer it's talking about but we never use...

S1: You have it? Isn't it like kind of expensive?

S2: We never use the silver part of it, so I don't know. And it's efficient with water. It saves water.

S3: Wait, how can you be using it without using the silver nanoparticles and then know whether or not it actually saves water?

This exchange took place during a unit that addressed the use of nanotechnology in a context familiar to these students. Goal 3 of the National Science Education Standards (NSES) calls for students to "engage intelligently in public discourse and debate about matters of scientific and technological concern" (NRC 1996, p. 13). The unit described in this chapter, "Clean—At What Cost?" (downloadable at *http://nano-cemms.illinois.edu/ssi*), focuses on addressing this goal through a series of activities that allow students to develop an understanding about the use of nanotechnology that has potentially direct effects on their lives. The unit was developed for middle and high school students for integration into the science curriculum at a variety of possible locations (such as in lessons on microbiology, properties of matter, or impact of science and technology).

Introduction

In order to "engage intelligently in public discourse" students need to be able to create and defend scientific explanations. This skill has been described as a key scientific practice (Michaels, Shouse, and Schweingruber 2008). We developed our materials to include the explanation framework of *claim*, *evidence*, and *reasoning* to provide the necessary support for students to develop this scientific skill (McNeill and Krajcik 2008; Novak, McNeill, and Krajcik 2009).

The unit opens with an introduction to products currently available that incorporate silver nanoparticles as an antimicrobial agent. Students are asked to research a product and present it to the whole class. Next, students conduct experiments testing the effects of silver nanoparticles on bacteria. This firsthand experience allows students to make connections between manufacturer's claims and their own experimental data. The concluding activity of the unit focuses on societal implications of this technology. Through a role-playing debate, students examine this technology from multiple perspectives.

These curriculum materials were developed by the Center for Nanoscale Chemical-Electrical-Mechanical Manufacturing Systems (Nano-CEMMS) at the University of Illinois. As a part of the center's goals, this National Science Foundation (NSF)-funded center is involved in educational outreach to prepare the next generation to deal with issues that result from advances in nanotechnology. The center brings together educators and researchers to collaborate and ensure that educational materials are engaging, meet state and national standards, and are scientifically valid.

Methodology

In this chapter, we report on the use of curriculum materials in two different schools. One school is a high school of approximately 900 students, located near a small midwestern urban community. The student body of this school is from diverse socioeconomic backgrounds. Students in two advanced placement biology classes and one general biology class participated in this study. The majority of these students were juniors or seniors in high school. The second school is a small private middle school of approximately 200 students within the same urban area. The students involved were members of a seventh-grade science class. Classroom data was collected and analyzed for trends in student learning connected to NSES goal 3. This data included pre- and posttests and classroom observations.

Description of the Curriculum Unit "Clean—At What Cost?"

Investigating Products

The unit begins with a whole-group discussion about a product that uses nanoparticles. A new washing machine, made by Samsung, washes clothes with water containing silver nanoparticles. As the clothes are being washed, the nanoparticles are deposited on the fabrics. As the clothes are being worn, the nanoparticles prevent bacteria from growing on the fabric. This should help reduce odors because it is primarily bacteria that produce the smell we associate with dirty clothes and body odor. This product was chosen specifically because it forms the basis of the societal implication piece later in the unit.

After this introduction, students are asked to identify and research another product that uses silver nanoparticles. Through the Project on Emerging Nanotechnologies website (*www.nano techproject.org/inventories/consumer*) students search a database of more than 1,000 nanotechnology products. The search will yield several hundred products that use silver nanoparticles. Students are asked to prepare a two- or three-minute introductory presentation on their product to the whole class. During these presentations, students discover that the silver nanoparticles are used as antimicrobial agents in a wide variety of products.

Testing Nanoparticles

After the presentations of the products, a class discussion is used to question whether silver nanoparticles are truly effective as antimicrobial agents. Students are asked to inquire how these claims might be investigated, which leads students into the second part of the unit.

To investigate these claims, students are introduced to basic concepts of microbiology in a contextualized manner. Techniques such as spreading of bacteria and incubation are explained and demonstrated. An agar plate is passed around so students can view and touch the agar. This experience allows students to gain a sense of what agar is and how much pressure can be applied when spreading the bacteria without tearing the agar. Students then begin the lab investigation to test for bacterial sensitivity to silver nanoparticles. (A more detailed description of this lab investigation can be found in Muskin et al. 2008.)

We have found that this laboratory investigation provides opportunities to have in-depth discussions about valid experimental designs. As students design their investigations, discuss what a control is and why it is a critical component of each of their experiments. After the data collection and analysis phase of the investigation, students are able to evaluate whether silver nanoparticles are effective antimicrobial agents. This evaluation leads them into the third part of the unit.

Debating Societal Implications

How technology affects society is often overlooked in science curricula. In this unit, this important topic is addressed with the culminating activity: a role-playing debate. Students are introduced to a problem scenario. Students assume different roles within this scenario, and to debate the problem, they research information relative to their roles. They then debate with other students representing different roles. Finally as a group, a recommendation is made based on the information and arguments.

The scenario we used featured a hospital that was considering the purchase of washing machines that use silver nanoparticles. This scenario is based on a real product that students were introduced to at the start of the unit. As the linens are washed, some of the nanoparticles settle on them and provide a measure of antimicrobial protection. In the scenario, the hospital is considering using this machine to help combat possible patient bacterial infections and to save money by heating the water less than a traditional washing method would.

Students are given one of five roles: health care worker, patient, hospital purchasing manager, hospital legal counsel, or environmental regulator. Each student receives information and a set of concerns unique to their roles. For example, students representing the health care workers are given statistics on hospital infections of patients by bacteria; those representing environmental regulators receive information about the effects of silver nanoparticles on the environment. Students meet with other classmates assigned the same role. These groups read and discuss the materials provided to them and identify research questions generated from their discussions. After researching their questions, students develop an understanding of the complexity of the issues facing someone in their position.

After students are able to articulate the position and their own
role, they are shuffled into new
groups. A jigsaw strategy is used
so that each new group has one
member representing each of the
five roles (Johnson, Johnson, and
Holubec 1998). If groups cannot
be made with one student per
role, additional students might
be assigned to the same role, or
a role might be omitted from a
group. These new groups represent a hospital advisory board. As
this advisory board, they need to
decide whether or not to recommend that the hospital adopt the
laundering method using silver
nanoparticles. Each student comes
into the discussion primarily aware
of his or her own role's position
and quickly learns that there are
other justifiable concerns. After
each team decides on their recommendation, they briefly present
their decision to the rest of the
class, discussing some of the issues
that arose and why they reached
the conclusion that they did.

We have found that this scenario works well with the roles provided; approximately half
of the groups recommend adopting the new laundry method, and the other half decide not to.
We felt it was important to find an issue where different group recommendations resulted to
reinforce the idea that not all issues clearly point to one outcome as better than another. After all
groups have presented their decisions, the class uses this activity as the context for a whole-class
discussion on how the risks and rewards of many issues do not point to a single correct decision.
This unit begins to show students that scientific knowledge is fluid, and attention needs to be
paid to how science is used.

Upon reflection, the authors noted that ways one teaches may unintentionally influence the
perspectives of their students. This bias may lead to a particular outcome with respect to the
debate. Teachers need to be aware of this potential bias so as to not influence students' discourse.
Although it is impossible to completely eliminate personal bias, care should be taken when
engaging in issues with multiple outcomes.

Results and Evidence of Successes

With post-assessment data, we have documented the learning that occurred as a result of student engagement in this unit. In examining students' responses, we identified three key themes linked to the NSES Goal 3. These themes are (1) developing awareness of different perspectives and potential bias, (2) the need for evidence to support a claim, and (3) the impact of a decision. In this section we highlight these themes by using student examples and articulate how they were developed during the enactment of the unit.

Theme 1: Developing Awareness of Different Perspectives and Potential Bias

This theme became evident as students analyzed different sources of data. This bias awareness allowed students to critically examine data they obtained for use as evidence.

> S1: I've learned that it can be hard to find completely objective information about scientific stories or reports from the popular press.

Students also realized that there were multiple viewpoints that needed to be considered.

> S2: It has shown me, although I already guessed, that different news sources show different sides of an argument.

> S3: It shows that there are many more sides concerning the story, i.e., many more factors than what the press usually reveals.

The majority of the students shared the understanding that these multiple viewpoints were warranted.

> S4: [The "Clean—At What Cost?" unit has] given me a broader view of the positives and negatives of scientific advances. The positive effect is that it replaces any conventional cleaners but can have negative effects on the environment.

Students learned that evidence could be used to support differing perspectives. One student demonstrated his understanding of this complexity by saying:

> S5: [The new washing machine] has positive and negative ideas associated with it. It seems to lower cost and increase efficiency but it also may damage the environment.

The students began to examine all sources of information and consider what bias the sources might have. Some students generalized this concept beyond the unit and stated that they now routinely look at the source of data as they consider its validity.

> S6: Now, when I hear scientific reports, I think about how in the nanotechnology unit we found that scientists who are trying to prove something might be a bit more biased, so I look at who wrote the scientific reports/stories.

Students realized that they needed to evaluate the source of the data and look for bias based on the source.

> S7:   It has affected how I think. I now see that reports can be influenced by personal views on a product and one needs to do a lot of research before forming an opinion.

One interesting result observed during the preparation for the debate was the manner in which students judged the validity of information. For example, one of the groups assigned the role of hospital manager was provided the website of the washing machine manufacturer. They quickly realized that the manufacturer might be a biased source and checked the validity of many of the claims made by the manufacturer. Other groups exhibited this behavior as well.

An additional example of students examining source bias was in the discussion of a middle school group assigned to the role of health care worker. While researching, this group found information contributed by an individual with a PhD. They assumed the information was accurate and unbiased. Later, they were surprised when they came across several sites with contradictory information to the first site. They then looked more closely at the authors of the sites and realized that the first site was hosted by an activist organization. They were amazed that they had been so strongly taken in by the first site. This group of students became much more careful in looking for possible bias.

Theme 2: Need for Evidence to Support a Claim

The unit highlighted the use of scientific explanations incorporating claim, evidence, and reasoning. This emphasis of scientific explanations was a component of all activities in this unit. The use of scientific explanations by students emphasized that they need evidence to support their ideas. It became apparent that as the students prepared for the debate, they began to develop a better understanding of why they needed to have data to use as evidence to back up their claims. Students saw that without evidence, unsupported claims can easily (and wrongly) be made.

> S8:   The results can be used to tell people that what they hear is not always true.

Students viewed evidence as a way of supporting a claim in their debate. The use of evidence prevented the discussion from becoming a "he said, she said" debate. Instead, it allowed students to focus on supported arguments and have a productive debate. This level of understanding is reflected by the student's statement about the need of evidence to support a claim and letting the audience decide if the evidence was sufficient.

> S9:   Well, you should show your results and state your findings. Maybe then you can change some minds.

Students realized that the use of evidence goes beyond just trying to change someone's opinions. Instead, evidence was used to evaluate the claim and its supporting evidence in order to formulate a reasoned group recommendation.

S10: It [evidence] tells about how they [silver nanoparticles] affect the environment and people. We can now decide whether or not to use silver nanoparticles.

Theme 3: Impact of a Decision

Through engaging in this activity, students realize that all decisions have impacts and that these impacts might not be good or bad but rather contain elements of both. In addition, students start to realize that through solving one problem, another is often created. A student made this observation in the statement below:

S11: With our results from the activity, you can see that some of the ways we fixed problems with the silver created new problems that science would have to fix.

Students realized that there were conflicting impacts that needed to be taken into account in any final decision. Students articulated an understating that some views focused on the good of the patient and not the good of the environment and vice versa. All of these different concerns needed to be considered in the final decision-making process. One student articulated the conflict when she wrote:

S12: Nanotechnology can change everyday lives. In the washing machine example, sterilized clothes were beneficial to the people but not necessarily to the environment. Nanotechnology had both good (clean/healthy stuff) and bad (harming good things in nature) qualities just like everything else.

Some students went further in their understanding and recognized that often we do not have enough evidence to know all the positive and negative effects.

S13: It has made me realize that though we try to find the answers to everything, we may not always have enough information to do so. Sometimes we just have to make smart, educated guesses and be happy with how that turns out.

Conclusions

As a result of this unit, students engage critically in a discussion about issues that potentially affect them, the community in which they live, and the larger society, as called for in the NSES. As curriculum developers, we believe that to engage students in scientific discourse around an issue with societal implications, it is critical to select a genuine context and use an inquiry-based approach in the design of the materials.

In developing this unit, we selected an issue involving nanotechnology to serve as the real-world context. The activities were designed to allow students to develop a scientific understanding of the issue through multiple inquiries. This type of curricular approach has been shown in the literature to improve student attitudes and their learning of science (Bennett, Lubben, and Hogarth 2007).

The sequence of activities allowed students to engage in multiple inquiries into the applications of silver nanoparticles. In this unit there was a range of activities that were student-directed and that engaged students in multiple scientific practices. These practices included designing and conducting an investigation, analyzing data, developing scientific explanations, conducting scientific research, and participating in scientific debate.

Nanotechnology lends itself to developing materials focused on interesting societal issues because of it only recently emerging in science and because of the promise it holds for the future as noted by Gardner, Jones, and Falvo (2009). Our collaboration with the center's researchers allowed us to be cognizant of both the positive and negative potentials of the use of these emerging technologies. These new technologies are often found at the center of societal issues and should be considered as rich contexts for future curriculum development.

Additional resources for this unit are available at *http://nano-cemms.illinois.edu/education* and at *http://nano-cemms.illinois.edu/materials*.

Acknowledgments

This work was funded primarily by the National Science Foundation under NSF Award #0749028 (CMMI). The opinions expressed herein are those of the authors and not necessarily those of the NSF or the University of Illinois. In addition, we would like to acknowledge Matt Ragusa for his invaluable assistance in the development of the "Clean—At What Cost?" unit and chapter.

References

Bennett, J., F. Lubben, and S. Hogarth. 2007. Bringing science to life: A synthesis of the research evidence on the effects of content-based and STS approaches to science teaching. *Science Education* 91 (3): 347–370.

Gardner, G., M. Jones, and M. Falvo. 2009. "New science" and societal issues. *Science Teacher* 76 (7): 49–53.

Johnson, D. W., R. T. Johnson, and E. J. Holubec. 1998. *Cooperation in the classroom*. 7th ed. Edina, MN: Interaction Book Company.

McNeill, K., and J. Krajcik. 2008. Assessing middle school students' content knowledge and reasoning through written scientific explanations. In *Assessing science learning: Perspectives from research and practice*, eds. J. Coffey, R. Douglas, and C. Stearns, 101–116. Arlington, VA: NSTA Press.

Michaels, S., A. W. Shouse, and H. A. Schweingruber. 2008. *Ready, set, science! Putting research to work in K–8 science classroom*. Washington, DC: National Academies Press.

Muskin, J., J. Wattnem, M. Ragusa, and B. Hug. 2008. Real science or marketing hype? *Science Teacher* 74 (4): 57–61.

National Research Council (NRC). 1996. *National science education standards*. Washington, DC: National Academies Press.

Novak, A., K. McNeill, and J. Krajcik. 2009. Helping students write scientific explanations. *Science Scope* 33 (1): 54–56.

Tahoma Outdoor Academy:

Learning About Science and the Environment Inside and Outside the Classroom

Oksana Bartosh
Canadian Council on Learning, Canada

Amy E. Ryken
University of Puget Sound, USA

Margaret Tudor
Pacific Education Institute, USA

Jolie Mayer-Smith
University of British Columbia, Canada

Setting

We encounter scientific information and products of scientific inquiry in our everyday lives, thus science literacy is a necessity (AAAS 1993; NRC 1996). Learning standards emphasize preparing students to become responsible citizens who are able to actively participate in debates and problem solving by using science concepts and skills. As the number of environmental problems the world faces grows, it is important for students to understand the science concepts and social issues underlying these problems and to be able to develop effective solutions. A promising educational approach involves using the environment as an integrating context to help students develop understanding of science as well as a sense of responsibility and stewardship toward the planet. Helping students develop another type of literacy—environmental literacy—is crucial for the 21st century (Hungerford 2010; NAAEE 2004).

Introduction

Science is often regarded as the most appropriate subject with which to integrate environmental education because scientific concepts form the foundation of environmental education (Ham, Rellergert-Taylor, and Krumpe 1988; Simmons 1989). The essential features of scientific inquiry, such as posing questions, organizing evidence to make a claim, and communicating results

(Martin-Hansen 2002) allow students to apply the "processes of science" (NRC 1996, p. 105) to complex environmental issues and interactions. Environmental education can provide novel and interesting learning contexts, activities, and perspectives. It also can make the science curriculum more appropriate and accessible for a wider range of students. Much of typical science education focuses on knowledge acquisition and learning concepts that are outside the experience of many students and therefore lacks relevance in their lives. In contrast, environmental education focuses on the local environment and community issues, making these alien science concepts more interesting and likely to be remembered (Dillon and Scott 2002; Gough 2002). It also can make science more culturally and socially relevant and help students develop novel and useful social and cognitive skills.

Although science is considered as the best fit for environmental education, proponents of infusion and interdisciplinary models recommend incorporating environmental education into a wide range of school subjects, including science, social studies, mathematics, language arts, and art. For instance, Rosenthal (2003, p. 154) advocates an arts-based approach; through this approach teachers can "cultivate systemic thinking, interdisciplinary problem solving, collaboration, and social and environmental responsibility" which is beneficial to all students entering "a world that demands creative and far-reaching responses to the damage we have wrought upon human and non-human systems."

Taking students outside of the classroom is recognized as a valid and important pedagogical practice; however, teachers at all levels feel challenged when faced with integrating informal, outdoor, and community settings with subject-specific curricula and mandated learning outcomes (Falk and Balling 2001; Michie 1998; Price and Hein 1991; Simmons 1998; Smith and Williams 1999). There is a need for models of integration that also address engaging students in field investigations and civic participation.

In this chapter we describe a high school program that uses the environment as an integrating context for science, language arts, health and fitness, and service-learning. We explore how the program influences students' understanding of science, environmental concepts, and inquiry skills.

Program Description: High School Outdoor Academy

The Tahoma Outdoor Academy is a program implemented in a high school in a suburban community of 17,800, 30 minutes outside Seattle, Washington. Over the years, teachers and staff have worked closely with the Pacific Education Institute (PEI)[1] to design interesting and innovative programs with inquiry, critical thinking, and environmental education as threads that unite different grade levels and subject areas. Aided by these coordinated efforts, the district has become one of the 10 highest performing districts in the state, with students scoring in the top 10% on the state standardized tests in mathematics, language arts, and science.

---

1. The Pacific Education Institute (*www.pacificeducationinstitute.org*) is a consortium of stakeholders interested in advancing student learning through curriculum designed around using the environment as an integrating context. It includes representatives from the business community, nonprofit educational and environmental organizations, state agencies, national environmental education programs, residential environmental learning centers, school districts, and individual schools.

Thematically organized instruction helps to integrate the normally disconnected course offerings found in comprehensive high schools and creates learning communities of teachers and students (NRC and Institute of Medicine 2004). The Tahoma Outdoor Academy Program is a yearlong, Grade 10 outdoor and environmental course that connects the study of science to other school subjects (including language arts, health, and fitness) as suggested by NSES Program Standard B (NRC 1996). The goals of the program are (1) to improve student learning; (2) to teach students to become environmentally literate citizens and lifelong learners; (3) to create a link between learning in school and the world beyond school; and (4) to explore local areas and the impact of humans on the environment. A service-learning component engages students outside the classroom in county nature parks and is consistent with NSES Program Standard D: "Good science programs require access to the world beyond the classroom" (NRC 1996, p. 218). Students conduct field investigations. Exploring the park environments, they "formulate questions and devise ways to answer them, they collect data and decide how to represent it, they organize data to generate knowledge" (NRC 1996, p. 33).

The program is taught by three teachers who integrate their subject areas with topics of environmental education, outdoor education, and stewardship. It enrolls approximately 90 students annually and includes students at all levels of achievement.

Students meet every other day (the school has a rotating schedule) and participate in the outdoor academy for an entire school day. A typical day starts with a whole-group session in which teachers and students discuss environmental issues using presentations, newspaper articles, books, and guest lectures given by community environmental advocates and wildlife biologists. Students then have time to discuss the issues with their peers and write reflections in their science journals. After the whole-group session, the class is divided into three groups of 30 students, one group per teacher. The groups rotate among the teachers for the rest of the day, and all students participate in one science, one language arts, and one health and fitness class. Lessons are often built around a common topic and use integrated assessments that require application of science, language arts, and fitness skills and concepts.

Outdoor Academy Curriculum

The Outdoor Academy curriculum includes themes of stewardship, environmental sustainability, and responsibility. It was designed by the three teachers, who tailored existing resources on outdoor and environmental education and ecology to the needs of their students (see Table 7.1, pp. 96–97).

The year begins with a water unit (September to December). Students explore their connections with the natural world through literature, scientific investigations, and such recreational activities as fishing and hiking. In the language arts class students read books about rivers and oceans and their roles in human lives (*A River Runs Through It* and *The Old Man and the Sea*). In science class they study aquatic ecosystems, visit a local river, and complete a field investigation research project exploring one aspect of the river's health. In health and fitness class students learn fly fishing skills, fish identification and classification, and discuss how to act responsibly in the outdoor environment.

In the second unit (January to March), students explore human behavior in the natural world. They discuss topics of risk taking, self-control, and perseverance. They read books about

Table 7.1. Outdoor Academy Program: Curriculum Topics (italics) and Activities (bullets)

	Language Arts	Health and Fitness	
September – December			
Topic: Humans and Nature. Exploring humankinds' relationship with the natural world. Personification of nature. Integrated Assignment: Research project on water/river issues	*Old Man and the Sea* • Characterization; metaphor; symbolism; allegory; allusion; theme/motif *River Runs Through It* • Personification (reinforced with Emmons and Green River trips) • Foreshadowing • Expository essay *River Packet* • Supplements the two novels • Reinforces learning targets above • Stewardship *Research component (river project)* • Working with secondary sources	*Fishing skills and activities* • Fishing equipment • Casting, fly tying, knots • Casting competition • Personification of rod, cast *Stewardship* • "Leave No Trace" *Reading water* *Life cycle of aquatic invertebrates* *Fish identification* *River practice*	
January – March			
Topic: World of life. Risk. Perseverance. Integrated Assignment: writing assignment	*Lord of the Flies* • Theme/Motif • Symbolism • Allegory • Diction *Thematic unit (perseverance, ambition, dual nature, addiction to risk)* • Into Thin Air • Touching the Void • Shackleton • other	*Orienteering* • LOTF exercise • GPS *Rock Climbing* • Terminology and equipment; techniques; rescue; safety; communication; movement principles; climbing activities *Fitness Testing* • Personal Fitness Plan *Touching the Void* • Large discussion	
April			
Topic: Biking and poetry	*Literature Circles* • Readings about nature and environment *Poetry* • WASL practice	*Biking* • Safety; maintenance; skills; field experience *Fitness Testing* • Personal Fitness Plan	
May – June			
Topic: Field investigations Integrated Assignment: "I went to the Woods" - culminating project with stewardship focus	*Into the Wild* • Discussion skills; diction; irony *Walden* • Stewardship • Text-to-text connections • Self-reliance and nature *"I went to the Woods"* • Personal narrative	*Stewardship* *River restoration* *Conservancy exploration*	

Science	Service-Learning	Other Field Experiences
Aquatic Invertebrates • Life cycles • Classification • Water quality (including chemistry, biology, environmental impact) *Experimental design (research project)* *Communities and Ecosystems* *Managing Human Affected Ecosystems* *Symbolism* • Spotted Owl; Salmon *River Culminating Project*	*Aquatic Invertebrate Survey* *Stewardship* • Identifying concerns/issues • Begin reconciling *Habitat Restoration* • Invasive species removal *Trail Building* *Links to curriculum* • Personification work • Theme work • Symbolism work • Reflective writing • Casting practice	*Hikes* • Green River • Mid Fork of Snoqualmie • Emmons/Glasier Basin • Holder/Carey Creek hike *Habits of Mind* • Gathering data through the senses • Responding with wonderment and awe • Thinking and communicating with clarity and precision • Listening with understanding and empathy
Organisms and Systems • Interrelationships • Limiting factors • Human/ Environment interaction • Continuity of life *Stream/River Dynamics* *Cartography/Mapping* • Soil investigations *DNA* *Research project*	*Data collection* • Erosion evaluation • Soil data collection • Mapping the area *Habitat Restoration* • Invasive species removal; planting of native species *Trail building* *Links to curriculum* • Orienteering • Rescue training	*Hike, Snowshoeing, etc.* • Vertical World • White River hike • Taylor Mt. Geocache hike *Amazing race* *Habits of mind* • Persistence • Taking responsible risks • Managing impulsivity *Other skills* • GIS training • Arcview training
Human Body Systems	*Habitat Restoration* • Trail building *Links to curriculum* • Outdoor poetry	*Biking* *Habits of Mind* • Thinking flexibly • Responding with wonderment and awe
Ecology *Identification of plants and animals* • Field guides • Systems • STELLA *Ethnobotany*	*Habitat Restoration* *Trail building* *Links to curriculum* • Filed survey/guide development • Outdoor reading/writing	*Hike* • Mailbox Peak • Overnight base camp

human behavior in extreme natural environments (*Touching the Void*), and learn rock climbing, mapping, GPS, and orienteering skills.

The third unit (April) focuses on poetry and recreation activities such as biking. Students explore local trails and read poems about the natural world.

The fourth, culminating unit (May to June) is designed around ecology concepts. Students read about and discuss ideas of survival in the wilderness, self-reliance, and self-actualization. In science they learn about local plants and native peoples' uses of them. The year ends with an overnight camping trip and a challenging hike on a local mountain, Mailbox Peak.

Environmental Education and Stewardship

Environmental education and stewardship are important components of the Outdoor Academy program. The curriculum uses these topics as a basis for integrating assignments and discussions. Environmental topics are included in discussions and presentations during the common hour in the morning. The following comment by a teacher illustrates the nature of program learning experiences.

> In large groups we usually talk about different issues. We talk about problems, current issues that appear in the newspapers; we talk about stewardship. So if something [is] going on, we usually use it as a focus for a discussion. We talk.

In addition, a number of field investigations, such as water quality and soil testing, are conducted throughout the year. In the first semester students do a group research project in which they investigate an issue related to a stream at one of the service-learning sites. Each student also completes a secondary research project in which he or she selects an environmental issue or a stewardship topic. Students are taught how to search for and work with information sources and how to write a report. One student explained the process:

> We wrote 30 or 40 different topics on the board and everybody took one they liked and we wrote tons of questions on it. We started off with global warming and then things that were affected by global warming and then we broadened it to this environmental stuff... you know, trees, fishing, life in oceans... things to do with the environment. Then we started to research them, and make note cards and write outlines. And now we are at the point when we are writing a paper right now. It's kind of the whole research process.

Finally, environmental and stewardship education is part of outdoor activities where students participate in service-learning projects. Again, emphasizing the importance of discussing environmental issues, the teacher explained, "It's something that we also do when we go outside. When we go on the field trips, we discuss how to be respectful and, you know, leave no trace. We talk about stuff like that a lot."

Service-Learning

Service-learning is a teaching and learning strategy that "integrates community service with academic study to enrich learning, teach civic responsibility, and strengthen communities"

(National Commission on Service-Learning 2002, p. 3). Some consider it to be one of the elements of a successful environmental education program (Athman and Monroe 2004). It allows educators to design programs that integrate academic learning with meaningful service that meets the needs of the community and includes personal development and stewardship. Teachers of the Outdoor Academy name "connecting" and "giving back to the community" as important goals of the program. As a result, two service-learning projects are designed and incorporated into the program. From September to June students work with county representatives on restoring the Log Cabin Reach natural area, removing invasive species and planting native plants. As one student described,

> We planted trees at Log Cabin Reach, we planted a lot, and then we removed blackberry bushes, because they are invasive and then we put this black mat over the trees that we just planted, so the blackberry bushes cannot grow back and kill the trees.

Another service-learning project involved trail building and maintenance in a neighboring nature park. A student shared, "We went to Taylor Mountain and we cleaned all the trails, we clean all the roots and stuff and the leaves, and that was interesting." Students visit the sites every two weeks for the whole day. In addition to the service-learning components, the sites are also used to teach classroom concepts and topics; for example, students conduct water and soil tests, practice fly fishing skills, write journal entries, and discuss "leave no trace" practices.

Curriculum Models: Field Investigations and Civic Participation

The Outdoor Academy program is guided by two curriculum models—one for field investigations and one for civic participation. As "science is not a fixed body of knowledge but an evolving attempt by humans to create a coherent description of the physical universe" (White 2003, p. 174), students learn how to conduct descriptive, comparative, and correlative field investigations to explore natural and scientific phenomena (Figure 7.1, p. 100). Three types of field investigations are explored. *Descriptive* field investigations involve describing or quantifying parts of a natural system. In *comparative* field investigations, data are collected on different populations of organisms, or under different conditions (e.g., times of year, locations), to make comparisons. *Correlative* field investigations involve measuring or observing two variables and searching for a pattern (Ryken et al. 2007; Windschitl et al. 2007).

The students' field investigations involve essential features of scientific inquiry (NSES Content Standard A), such as asking "a question about objects, organisms, and events in the environment," planning a systematic approach to data collection, and developing "descriptions, explanations, predictions, and models using evidence" (NRC 1996, pp. 122, 145). In addition, the investigations help students develop skills such as mapping and perspective-taking to study community sites (CEE 2002). Students talk vividly about their field investigations. One student said,

> We learned a lot about the experimental design and collecting data and …. procedures to find levels of DO [Dissolved oxygen] and that kind of stuff.. . We had to make a hypothesis

Figure 7.1. Sample Student Field Investigation Research Questions

Essential Question How does housing development in the community affect fish in the nearby stream?	
Field Investigation Type	**Investigative Questions**
Descriptive studies Choose measurable or observable variables.	What areas of the stream are more populated with fish? What do fish eat?
Comparative studies Choose one focus variable to be measured and observed in at least two different locations, times, or populations.	How does the level of dissolved oxygen needed by fish vary by season? Is there a difference in numbers of fish upstream and downstream from the housing development?
Correlation studies Choose two continuous variables to be measured together and tested for a relationship.	Do the numbers of fish decrease with more houses being built along the river? Is there a relationship between fish population decreases and human population increases?

about the water, like the plants, how they grow and we had to make a whole science experiment, and then we had to do it and collect all the data, and that was fun.

Students are able to talk at length about their investigations and connect them to the research of other scientists, as this student explained,

> We've done a lot with rivers and bugs in the rivers. First we wrote questions about the river and what would look like good scientific investigation questions. So I did some kind of bug study, how logs in the river affect bugs. So we had to go to the river and go to different sites and collect data and go into the river and get bugs… It went pretty good. … [I found] pretty much what I expected. I mean I thought that there would be more bugs downstream than it would be upstream from the logs … and it was pretty much what I thought. I did a lot of research in class before we went there, so I kind of based my study on that research. And it was right… I think because the water runs past the logs, and there is material, organic stuff that comes off of it which bugs eat, so there are more bugs downstream because all that stuff is washed off the log and they eat it… so downstream there is more food, so there are more bugs.

Consistent with the NSES Science and Technology Content Standard E and Science in Personal and Social Perspectives Content Standard F (NRC 1996), the Outdoor Academy uses benchmarks for civic participation (Figure 7.2) to emphasize that the scientific process involves engagement with stakeholders within the community (Ferguson et al. 2004). Students identify an issue that involves interplay between human and natural systems, consider the perspectives of multiple stakeholders, create a plan of action, and implement and evaluate the plan. One student reported,

We're doing a project right now; we chose something that we have to think about. I chose waste and I found out what they do to our environment. And everybody does his own topic. The teachers teach us about it and we are still finding out facts and then we'll have to write a paper about it. I'm writing about how we can recycle, and that we need to reduce the waste because if affects our environment. What we do now will affect future generations.

Figure 7.2. Civic Participation Integrated Benchmarks

Step 1: **Work with peers and community members to identify and describe a local, regional, or international issue that involves interplay between human and natural systems.** • Identify and describe the natural system, human political and economic systems, and human cultures.
Step 2: **Identify the major players and stakeholders (e.g., government agencies, diverse cultural groups, producers, consumers, organizations, and individuals), their perspectives, interests, and resulting positions.** • Determine and describe why the stakeholders hold their given values and beliefs.
Step 3: **Explore ways to address the issue by creating and/or considering alternative solutions related to the issue.** • Consider the feasibility, responsiveness, and likely effectiveness of different action plans.
Step 4: **Collaborate in the development of a plan of action in response to the issue.** • Include objectives, timelines, tasks, products, division of labor, and evaluation methods.
Step 5: **Prepare a rationale for the plan.** • Consider the impact on the major stakeholders, response of stakeholders, potential consequences (e.g., environmental, human health, public policy, and economic) of implementation.
Step 6: **Implement the plan.** • Collect evidence of civic behaviors (e.g., communication with stakeholders, videotapes, surveys or evaluations, photographs, products, and action logs).
Step 7: **Evaluate the effectiveness of the plan.** • Evaluate effectiveness based on initial criteria and describe the unintended consequences of implementation.

Assessment Strategies

To evaluate student learning, teachers use a variety of assessments that range from traditional quizzes and short reflections to large, integrated research and writing projects. For example, students design and carry out an investigation related to a water issue (science) during their outdoor classes and write a report summarizing their findings (science and language arts). They also write papers on environmental issues in which they have to demonstrate both understanding of the concepts related to the topic (science) and their ability to use and reference various sources of information (language arts). Another integrated assignment asks students to write a fictional mystery story: students are provided with the beginning of the story and the facts from a "crime scene" in the mountains. They create a story (language arts) that solves the mystery using knowledge of DNA concepts (science) and skills and knowledge of rock climbing (learned during their health and fitness classes). Finally, at the end of the year, students reflect on their experiences in the program through the "I went to the woods" writing project in which they explore their relationship with the natural world and their understanding of stewardship values.

Evidence of Student Learning

The Outdoor Academy program provides a model of how the environment can serve as an integrating context for learning the combined content of science, language arts, and health and fitness at the high school level. Inquiry tasks are used to document student learning, and students' reflections are used to understand their experiences in the program.

Inquiry Performance Tasks

Two scenario-based inquiry tasks developed by the Pacific Education Institute are used to assess student knowledge of science topics as well as their ability to design a field investigation. The tasks are administered at the beginning and the end of the year. At the beginning of the year, students complete a soil percolation task which asks them to design an experiment on how different locations affect water percolation time. At the end of the year, students complete a hot spot task that requires them to design an investigation to determine how different locations affect the surface temperature of the ground. Students identify and include the responding (or dependent), manipulated (or independent), and controlled variables. They also describe steps of the procedure that could be used to repeat the investigation successfully. Student work is evaluated using a rubric (Figure 7.3).

Analysis of inquiry performance tasks demonstrates that students develop better understanding of the field investigation process over time. Most students (75%) were able to design more elaborate and complex investigations to answer the proposed research questions. On the pre-program task, some students struggled with identifying manipulated (independent) and controlled variables; in contrast, on the post-program task the majority of the procedures included all three variables. At the end of the year, students also developed better tables for recording the measurements and more detailed step-by-step descriptions of the investigations.

Figure 7.4, page 104, illustrates a representative inquiry performance task response, in which the student makes a prediction ("the higher up the more it rains") and then gives a reason for her thinking ("because the higher up one goes the higher the humidity/moisture

Figure 7.3. Inquiry Performance Task Rubric

Prediction/Hypothesis	Student predicts how manipulated variable (e.g., time, location, organism, population) affects the responding variable. Students provide a reason for their prediction/hypothesis.
Materials	Student lists materials and tools needed to perform the investigation.
Controlled Variable (kept the same)	Student states or implies at least two ways that measuring variables and/or sampling are kept the same.
Manipulated Variable (changed)	Student states what is changed (e.g., location, substrate, habitat, time, organism, or population).
Responding Variable (measured)	Student states how data is observed, measured, and recorded.
Logical Steps With Trials Repeated	The steps of the procedure are detailed enough to repeat the procedure effectively. Student indicates that data will be recorded or creates a data table that includes date, time and weather conditions. Student notes that data will be measured more than once at each location.

is found in the air"). In addition, the student states three ways that sampling and measuring are kept the same (e.g., collect rain for three minutes, select sites that are open, and collect in the same rainy weather), notes three different elevations for study (manipulated variable), and includes multiple trials at each elevation.

Student Reflections

To explore the types of learning experienced as a result of the outdoor academy program, interviews were conducted with a subset of students at the beginning and the end of the year. The interviews demonstrated that students learned about environmental science concepts, interactions among humans and the environment, and social skills.

Understanding of Science Concepts

The interviews revealed that students developed understanding of science concepts and began to make connections between areas of science as well as connections between science and other subject areas. Students developed inquiry skills, learned how to conduct field investigations, and commented that learning became more engaging, relevant, and personal. The curriculum included physics, chemistry, and biology as well as environmental science topics, which assisted the Outdoor Academy students to develop an understanding of relational concepts such as populations, cycles, and carrying capacity. Many of these topics and issues were explored through hands-on investigations and group work. Students were encouraged to pose questions, search for information, and explore multiple perspectives and points of view. One student said,

> We asked some questions, and we had our own questions… we had to get some information for it. It could focus on water quality; it could ask how the soil affects the water quality, and then how the native plants in the area affect the water level. Then we had

Figure 7.4. Inquiry Performance Task: Student Response

Question: How does change in elevation up the west side of a Cascade mountain affect the amount of rain?

Hypothesis (prediction): If the change in elevation affects the amount of rain then the higher up the more rain you'll get because the higher up the more amount of humidity/moisture is in the air

Materials: • elevation counter on a watch

• measuring cup

• stop watch

• cascade mountain

Procedure: You may use this space for a labeled diagram to support your procedure.

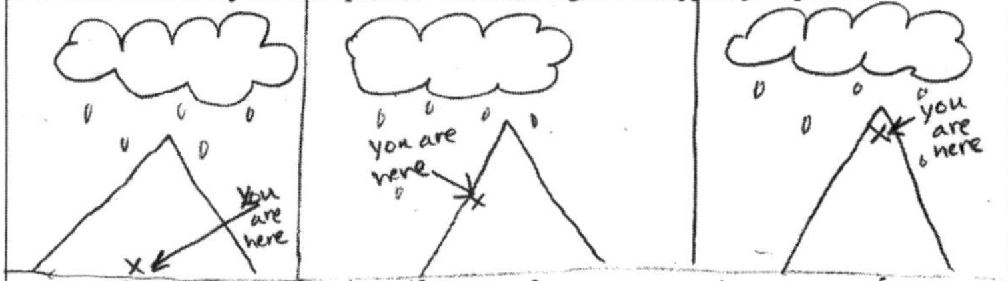

3 different elevations of a cascade mountain

Procedure (continued): 1. Record date, time and cascade mountain

2. Describe rain weather (sprinkles, hard rain, freezing rain, ect.)

3. Pick 3 different elevations on the mountain, where you will test rain fall amounts.

4. Record elevations (ex. 5ft, 500ft, 100ft.)

5. Go to each elevation, and pick a spot that is open and not crowded by trees.

6. Hold a measuring cup out and time for 3 minutes.

7. Record the amount of rain collected during the 3 minutes in mL.

8. Repeat steps 5-7 for the other two elevations.

9. Repeat the entire investigation 3 more times, using the same elevations and almost the same rainy weather.

a whole bunch of different kits we could use to go and find additional information. My question was [about] the difference between native and invasive species and the water quality. I found there were more native species where the water quality was better. Where there are more invasive species like raspberries, no blueberries that were not supposed to be there, the water quality was poor.

Field investigations were part of the program's curriculum, and students were required to design and conduct an experiment that dealt with a water issue. However, unlike many of their peers who explored similar topics by reading a textbook, Outdoor Academy students used their service-learning sites as areas for data collection. This made the process of learning more real, memorable, and interesting.

Because the Outdoor Academy program integrates several subjects, students were able to develop connections between subject areas as they explored environmental topics through the lenses of difference disciplines. They recalled studying creative and technical writing, symbolism, metaphors, allegories, and motifs in English, and learning games, fitness, and recreational skills in physical education. Students reflected that integration of the three subject areas made learning a more integrated experience. One said,

> In English we are learning to cite works and stuff like that, which is also connected to a project in our science class. It's helping... we were working on fly fishing and we study lots of fish in science, and then we work on the actual fly fishing in PE, and then we read a book about fly fishing in English. It's always like mixed all together. They are integrated.

This student's comments demonstrate that meaningful assignments increase student commitment to learning:

> Mr. V. had us write, which incorporated genetics and rock climbing and story writing, all in one assignment. That was pretty cool. It was a murder mystery, and you had to explain what happened, who did it [committed murder], how the detective found out who did it and why that person did it. It was actually pretty fun to write. I ended up with 5-6 pages more than what was required. He asked for 12 and I ended up with 19. So, I really liked doing it.

Learning About Environmental Issues

Over the course of the year, students conducted a number of research projects that focused on various environmental issues. Topics identified through group discussions included global warming, climate change, pollution, acid rain, and deforestation. NSES Content Standard F emphasizes the importance of students exploring how humans use natural resources and how human activities impact natural systems (NRC 1996). Student awareness of environmental issues increased, as one student noted,

> We picked an environmental issue that we liked, I picked deforestation and learned how it's affecting the world in general and what the problems are because of it right now... really important issues that we are learning about and we are more aware about.

Students learned how to search for and evaluate information, compare stakeholders' perspectives on environmental issues, and consider possible solutions, and reflect on how the topic each student selected was related to individual lifestyles and choices.

Learning About Interdependence Between Humans and the Natural World

Through discussions and readings, Outdoor Academy students developed an understanding of the interdependence of human and natural systems and explored various cultural perspectives on these issues. Students saw humans as an integral part of the environment and came to understand how human activities and choices affect the delicate balance of natural systems. For example, students learned that humans are a part of the environment and human actions have indirect impacts on the state of the environment in other parts of the world. A student commented,

> The environment it is not like way back in the woods where there are no humans, and everything that we do. That is what I've learned this year so far, that everything that we do affects things that we never even see. And down the road, there are a lot of things that we don't see, but we have to take them into consideration, even if we are not directly affected by something else.

Participation in the Outdoor Academy program encourages students to rethink their personal lifestyles and choices so they can consider how they might act responsibly toward the planet. Students reported making changes in their everyday actions, noting that their sense of responsibility had grown from experiential learning aspects that gave them numerous opportunities to enjoy the "great outdoors." The program changed students' behaviours and empowered them to take environmental actions, viewing environmental stewardship not as an obligation, but as a way to solve problems, as one student remarked,

> The care for the environment is not just, "take care because you have to," it's really, you want to do it. It's really trying to understand it, to learn about the environment, and stuff like that. Do not just think about it as an obligation, but to think of it as... people think about it as they need to do it. But you should think about it as what it is expected from everybody because of people you love. We talk about what it should look like. And when we go there, on field trips, before we went we got to learn what it is like properly act out there. We learn... to leave no trace, when we leave none of our garbage behind. We don't break stuff, we leave the environment as it was before we came and when we began it was just a role, but they don't need to tell us any more, we just do it, because we understand what the problem is. And what we can do about it.

Lifelong Learning Skills

Students in the Outdoor Academy collaborated and learned to engage in scientific conversations as recommended in the NSES Teaching Standards. Through group projects (inside and outside the classroom) and discussions and reflections on readings and experiences, students developed better understanding of themselves and their classmates. They developed communication and public speaking skills, and learned how to work independently and in groups. Through the various activities they gained insight on their life choices and personal goals. As one student's comments below demonstrate, they learned about their own beliefs and also the beliefs of others,

> I think I learned more about myself, what I want do, where I want go… being more outgoing, more comfortable with speaking in front of people, presenting, being myself. I learned about my perspective on a lot of different issues, politics and things like that. I learned a lot about what my beliefs are, how to handle myself in situations, how to push myself, to try harder and motivate myself. I learned a lot of things in science, basic science, but fun I guess. Language arts and English, I actually I learned more in that class probably than in the others, because all of the writing tests that we did were kind of like what would you change about yourself… questions you have to think a lot about, make you question things and just make you think a lot, and I learned a lot about myself, and how I feel about things. Then we read some of them aloud and kind of learned what other people's perspectives on all things that we've talked about. I learned how to be open minded to other people's ideas.

Students developed better group skills and learned how to work with peers, how to help each other, to be open-minded with other peoples' ideas, and to value others' opinions and perspectives. As this student said,

> I learned a lot of maturing things… how to be a part of a group, how to respect other peoples things, and … the way the people are. I learned to function well with other people who might not necessarily be the same as [me]. And then, I learned respect overall for everything, learning to be… instead of being loud and being in everybody's face all the time, to sit back and listen to what other people are saying and … I guess just be more receptive to other people.

The program provides many physical, mental, and intellectual challenges for students. The large, integrated research projects require that students apply knowledge and skills from all three subject areas, and go beyond the regular high school curriculum. In addition to curriculum challenges, the program includes a lot of intense physical activities, such as tree planting, removal of invasive plants, hikes, mountain biking, and rock climbing. For many students, especially those without prior outdoor or recreational experience, the level of difficulty was challenging. Students however, emphasized that the program taught them perseverance and how to face these challenges, as one said,

> They [the teachers] don't tell you how to make it up to the top… you can just sit down and wait for them to come back down when they are on the hike. The challenge is to keep going like, no, I am making it to the top. Because they gave the option of not doing it or doing it. And you want to show them you can. I am not in this class just because I thought it was easy, I am in this class because I am going do what we are going to do. So challenges really

... because they are not there to push you anymore, you have to push yourself. And the challenge is keep on pushing yourself to make it better. And that's one of the hardest things.

Students particularly enjoy the social aspects of the program, the hands-on learning, the encouragement from teachers who created fun and interesting lessons, and the fact that the program activities often had real-life applications (such as collecting research data for county officials regarding water and soil quality). These activities kept students motivated, as one noted,

Personally for me school, especially at this time of the year, during the last couple of months, It's hard to stay focused a lot of the times. And stay motivated ... and this class helps you... It's almost easier to stay focused. It's easier to just get your work done and do a lot of stuff because all these things, you enjoy them I guess.

Conclusion

The Outdoor Academy program illustrates how using the environment as an integrating context for learning at the high school level can help students develop an understanding of science concepts; engage them in debates about local, regional, and global issues; and involve students in community projects. The interdisciplinary nature of the Tahoma Outdoor Academy program helps students make connections among topics and phenomena in science, technology, the environment, and society, and makes school learning relevant and engaging. Intellectually challenging discussions about society and the environment, reflections on the program experiences, whole-day activities outside the classroom, and active engagement with community groups, provide students firsthand experiences to engage in community life and become active and responsible citizens.

Acknowledgments

Working on this study was an exciting journey, and we would like to thank Nancy Skerritt, Assistant Superintendent, Director of Teaching and Learning at the Tahoma School District and our colleagues at the Pacific Education Institute for their constant support, guidance, and advice. Our special gratitude goes to the teachers and students of the Tahoma Outdoor Academy who participated in the study and who shared with us their exciting experiences and ideas. Without them this study would not be possible.

References

American Association for the Advancement of Science (AAAS). 1993. *Benchmarks for science literacy*. New York: Oxford University Press.

Athman, J. A., and M. C. Monroe. 2001. Elements of effective environmental education programs. In *Defining best practices in fishing, boating, and stewardship education*, ed. A. J. Fedler, 37–48. Washington, DC: Recreational Boating and Fishing Foundation.

Council for Environmental Education (CEE). 2002. *Science and civics: Sustaining wildlife*. Houston, TX: Council for Environmental Education.

Dillon, J., and W. Scott. 2002. Editorial: Perspectives on environmental education-related research in science education. *International Journal of Science Education* 24 (11): 1111–1117.

Falk, J., and J. Balling. 2001. The field trip milieu: Learning and behavior as a function of contextual events. *Journal of Environmental Education* 76 (1): 22–28.

Ferguson, L., M. Tudor, O. Bartosh, and T. Angell. 2004. Technical report #1: Environmental education frameworks that integrate Washington State's learning standards. Olympia, WA: Pacific Education Institute.

Gough, A. 2002. Mutualism: A different agenda for environmental and science education. *International Journal of Science Education* 24 (11): 1201–1216.

Ham, S. H., M. H. Rellergert-Taylor, and E. E. Krumpe. 1988. Reducing barriers to environmental education. *Journal of Environmental Education* 19 (2): 17–24.

Hungerford, H. R. 2010. Environmental education (EE) for the 21st century: Where have we been? Where are we now? Where are we headed? *The Journal of Environmental Education* 41 (1): 1–6.

Martin-Hansen, L. 2002. Defining inquiry: Exploring the many types of inquiry in the science classroom. *The Science Teacher* 69 (2): 34–37.

Michie, M. 1998. Factors influencing secondary science teachers to organize and conduct field trips. *Australian Science Teacher's Journal* 44 (4): 43–50.

National Commission on Service-Learning. 2002. *Learning in deed: The power of service-learning for American schools. www.wkkf.org/pubs/PhilVol/Pub3679.pdf*

National Research Council (NRC). 1996. *National science education standards.* Washington, DC: National Academies Press.

National Research Council and Institute of Medicine. 2004. *Engaging schools: Fostering high school students' motivation to learn.* Washington, DC: National Academies Press.

North American Association for Environmental Education (NAAEE). 2004. *Excellence in environmental education—Guidelines for learning (PreK–12). www.naaee.org/programs-and-initiatives/guidelines-for-excellence/materials-guidelines/learner-guidelines*

Price, S., and G. Hein. 1991. More than a field trip: Science programs for elementary school groups at museums. *International Journal of Science Education* 13 (5): 505–519.

Rosenthal, A. 2003. Teaching systems thinking and practice through environmental education. *Ethics and the Environment* 8 (1): 153–168.

Ryken, A. E., P. Otto, K. Pritchard, and K. Owens. 2007. *Field investigations: Using outdoor environments to foster student learning of scientific processes.* Olympia, WA: Pacific Education Institute.

Simmons, D. 1998. Using natural settings for environmental education: Perceived benefits and barriers. *Journal of Environmental Education* 29 (3): 23–32.

Simmons, D. A. 1989. More infusion confusion: A look at environmental education curriculum materials. *Journal of Environmental Education* 20 (4): 15–18.

Smith, G. A., and D. R. Williams, eds. 1999. *Ecological education in action: On weaving education, culture and the environment.* Albany, NY: State University of New York Press.

White, R. 2003. Changing the script for science teaching. In *A vision for science education: Responding to the work of Peter Fensham*, ed. R. Cross, 170–183. London: RoutledgeFalmer.

Windschitl, M., K. Dvornich A. E. Ryken, M. Tudor, and G. Koehler. 2007. A comparative model of field investigations: Aligning school science inquiry with the practices of contemporary science. *School Science and Mathematics* 1 (107): 367–390.

Developing Students' Sense of Purpose With a Driving Question Board

Ayelet Weizman, Yael Shwartz and David Fortus
Weizmann Institute of Science

Setting

We describe an instructional tool called a Driving Question Board (DQB) that can support students' sense of purpose by helping them make connections between their personal life and interests and the focal science and technology content of a project-based unit. The DQB is a large poster-board presenting the driving question surrounded by sub-questions raised by the students. It is established by the class as a group, representing the knowledge constructed by the classroom community throughout the unit. The DQB serves as a visual organizer, provides a synthesis of the various contextualizing features in the unit, and helps students find their way through the unit as with a road map. We follow previous studies to define "sense of purpose" in project-based learning using three characteristics: making explicit connections between the various activities in the unit and the main learning goals; focusing students' thinking, and breaking big questions into smaller ones.

In this chapter we report on the pilot of a sixth-grade project-based unit on light at three different schools—urban, suburban, and rural—and follow the use of the DQB by students and teachers at three sites. Pre- and posttests, individual interviews, and classroom videos were used to evaluate students' learning and their attitudes toward the unit in general and the DQB in particular.

Introduction

Research indicates that students often do not see science as relevant to their lives outside of class. Therefore, the National Science Education Standards (NRC 1996) recommend contextualizing curriculum materials in science by relating science to students' personal and social lives. Introducing scientific concepts in authentic and relevant contexts makes science meaningful, enhances intrinsic motivation, and fosters student learning (Blumenfeld and Krajcik 2006; Blumenfeld et al. 1991; Deci and Ryan 1987; Edelson, Gordin, and Pea 1999).

Contextualizing instruction may help students make sense of complex scientific ideas because the use of meaningful problems or situations provides students with a cognitive framework of more available links they can connect to or "anchor" their knowledge (Kozma 1991). The prior ideas that students bring to the science classroom are linked and integrated with new ideas, and reorganized to structure understandings of scientific phenomena (Linn 2000). By contextualizing instruction the teacher or educator can focus students' attention on the connections among concepts, and help organize and integrate students' knowledge by engaging them with scientific ideas from multiple perspectives (Rivet 2003). Thus contextualized instruction puts more emphasis on understanding scientific concepts, connections, and processes, instead of emphasizing knowledge concerning facts and other information.

The National Research Council specifies scientific inquiry as the recommended teaching approach, emphasizing that students should investigate phenomena relevant to them personally. They specify that curriculum should be inquiry-based in order to introduce students to scientific practices that represent the disciplinary norms of scientists as they construct, evaluate, communicate, and reason with scientific knowledge (NRC 1996). Inquiry-based instruction presents many challenges to teachers. Research indicates that teachers

1. rarely connect the activity to the learning goals or even to the main question that drives the investigation;

2. content is sometimes overlooked, with minimal discussion time devoted to scientific ideas;

3. often use the goal of completing school tasks as the "need to know" for the activities;

4. rarely discuss the reason for the activities; and lastly,

5. provide vague feedback to students regarding their academic work ((Harris 2005).

Another challenge for teachers has to do with the fact that students in diverse classes have different experiences with science and different prior knowledge. Teachers need to have access to tools in the classroom that will provide students the opportunity to build on their personal experiences and at the same time create a common base for "science talk" that will illustrate the class as a community of learners (Wenger 1998; Shwartz, Weizman, and Fortus 2009).

One response to these challenges and opportunities is Project-Based Science (Blumenfeld and Krajcik 2006), that employs contextualized instruction in four ways: (1) It uses problems and situations with implications out of school and that are meaningful to students (Edelson et al. 1999); (2) The meaningful problem situation motivates students to understand the content and engage in the task of science learning, and provides a purpose for knowing science ideas and concepts (Krajcik et al. 2002); (3) Using some form of anchoring situation engages students with the scientific content, providing a common experience for all students; and (4) Engaging with the same problem over an extended period of time allows for analysis from multiple perspectives (Marx et al. 1997).

Contextualizing the inquiry process helps students value the usefulness and plausibility of scientific ideas. All Project-Based Science units use a driving question (Krajcik et al. 1998) to create a context for learning. A driving question is a rich and open-ended question that uses everyday language to make connections with students' authentic interests and curiosities. The question cannot be solved immediately, but requires a series of investigations that can be accomplished in or around the classroom during several weeks.

However, the open-ended characteristic of the driving question creates several challenges including: (1) The solution usually involves multiple sub-questions, some which may seem somewhat unconnected to the others; (2) Students will likely have their own questions related to the driving question, that they would like to investigate; (3) Students may have difficulty recognizing the connection between the relevant science principles that are needed to create a response to the driving question; and (4) Because of the time required to develop a full explanation to the driving question, students may lose track of where they are. Because of these challenges, the curriculum should provide scaffolds to facilitate students' learning and help them recognize connections among activities related to the driving question. Such scaffolds include sub-questions that break the driving question into smaller ideas, and presentation of evidence collected and artifacts generated in earlier stages of the unit.

A tool that was developed to organize these scaffolds is called a Driving Question Board (DQB) (Singer et al. 2000). The DQB is a large poster used to collect, sort and present the sub-questions of the driving question and their answers, as well as related products created by the students.

The sub-questions are generated by the students mainly during the first lesson of the unit, but can be modified later. The driving question is written on the DQB, surrounded by the sub-questions that the students pose, and is dealt with in the unit. The process of taking a big, complex question and breaking it down into manageable sub-questions scaffolds students' problem-solving skills.

The DQB serves as a visual reminder and an organizational tool for the various scientific concepts, both for teachers and students. It helps organize learning and focus on specific content topics, by showing where the class has been, what students have learned and done on the way, and where they will be going. It also explicitly shows the integration of concepts and processes since it presents links between them in the form of questions.

At the end of each lesson, the teacher revisits the DQB for questions that have been answered during the lesson and to add questions that may have arisen. This allows students to draw upon previous knowledge and make connections between prior knowledge and newly learned concepts. The DQB provides explicit connections between the various activities and the driving question, supporting the context created by the driving question.

In addition, artifacts generated during investigations are presented on the DQB. Thus it is constructed by the students as a community of learners. Since it is created by the students, it can provide them with a sense of ownership.

The advantages of the DQB are especially prominent in inquiry-based learning, since it is helpful for integrating all aspects of science content; it literally shows the connection between activities, questions, and explanations; it helps to keep track of long-term investigations and reminds students about the context of individual activities and how they are related to the driving question.

Purpose of the Study

Our main goal was to investigate the influence of the Driving Question Board, as representing all the contextualizing tools in the unit, on students' learning. We looked for evidence for the influence of the DQB on the development of students' sense of purpose, which we characterize by their ability to: (1) make connections among science concepts, the unit learning goals and the driving question, and then relate them to their personal and social lives; (2) feel focused and organized in their learning; and (3) solve small questions derived from the big driving question. The research questions we pursued were: (1) How was the DQB used in the different classrooms? (2) What was the influence of the DQB on students' development of a sense of purpose, as defined above? (3) Did students find the DQB helpful?

Context

The Curriculum and Its Relation to National Standards

The context for the study is a Project-Based Science unit on the nature of light, in a three-year middle school reform-based curriculum called Investigating and Questioning Our World Through Science and Technology (IQWST). The curriculum is standards-based and learning-goals driven, based on the backwards design approach originally suggested by Wiggins and McTigte (1998). It includes units in four science strands: physics, chemistry, biology, and Earth science, for grades 6–8. Each year includes a sequence of four units, integrating the concepts with scientific practices such as modeling, evidence-based explanations, and design of investigations. The coordinated curriculum is an attempt to address the recommendations for developing coherent understanding of core scientific ideas as well as supporting inquiry-driven and contextualized learning (Shwartz et al. 2008).

Both the content and the practices' learning goals for each unit are standards-based, and were carefully chosen, specified, and elaborated (Krajcik, McNeill, and Reiser 2008). Then, a relevant, real-life and motivating context for learning these goals was chosen. Activities and readings were selected and designed to be motivating and have personal relevance and importance for students. To link scientific practices to content, we defined what students should be able to *do* with the knowledge rather than simply defining what they *know*. The level of inquiry varies from guided inquiry in the first units to a more and more open inquiry as the students gain expertise with the practices.

The curriculum addresses all eight categories of learning science as specified by the National Standards with particular attention to unifying concepts and processes in science, science as inquiry, and science in personal and social perspectives (NRC 1996, p. 6). Based on research and preliminary experience with inquiry-based units (Krajcik et al. 1998; Singer et al. 2000) IQWST units focus on four scientific practices: (1) Modeling and using models in science; (2) Data Gathering, Organization, and Analysis (DGOA); (3) Constructing evidence-based explanations; and (4) Designing investigations.

Practices specify how students should be able to *use* knowledge in meaningful ways, rather than what they should *know*. A scientific practice involves both the performance of scientific work and understanding the underlying meta-knowledge that articulates why the practice takes the form that it does. Learning these practices is essential for students to understand science as a way of knowing and not just as a body of facts. However, little careful planning has taken place regarding how to help learners develop these practices over time (Pellegrino, Chudowsky, and Glaser 2001). Learning these practices does not develop instantaneously, nor arise as a result of a single exposure. Rather, learning such complex practices takes time and numerous carefully scaffolded experiences. Our working hypothesis is that with a careful design and support, both for teachers and students, students will be able to apply complex scientific practices at a relatively high level over the middle grade years. The design of the instructional materials included the development of learning progressions that support students developing deeper understandings of scientific inquiry practices (Fortus et al. 2006).

Contextualizing Features in the Curriculum

In order to contextualize inquiry, IQWST materials are organized around a specific relevant context, which drives the learning of the content concepts. Each unit is organized around a rich open-ended question (the driving question) and must meet the following criteria: Is it worthwhile? Is it feasible? Is it real-word? Is it meaningful? The driving question interests and motivates students. Moreover, the science concepts and skills are keys to understanding and answering the driving question. Designing the units around a relevant driving question makes the context a central thread of the process of learning science, in contrast to more traditional approaches, which introduce the relevancy of the scientific idea at the end of the unit, or as enrichment reading assignments. Some examples of IQWST driving questions are: How can we smell things from a distance? Why do some things stop while others keep going? Where have all the creatures gone? How does water shape our world?

Discussions as Means for Preparing Students to Engage in Public Discourse

Following the NSES recommendation to engage students intelligently in public discourse and debate about matters of scientific and technological concern, the curriculum provides support and scaffolds for discussions. Support for teachers include an introduction about discussions in the teacher guide, integration of discussions in different places in each lesson, indication of the type of each discussion (including brainstorming, synthesizing, summarizing, and pressing for understanding), and suggestions for relevant prompts for each discussion. Every lesson includes several discussions accompanying the investigations and activities. These are discussions integrating content with practices and helping students process and express their thinking (Shwartz et al. 2009). In another study (Weizman et al. 2008) we found that whole-class discussions contribute to students' construction, communication, and evaluation of models in the same unit that is investigated in the current study. Here is an example from the curriculum:

Brainstorming discussion of the anchoring activity

Teacher Background Knowledge:
In IQWST, the purpose of a brainstorming discussion is to share ideas without evaluating their validity. See details on the "Classroom Discourse" page in the front materials. The following prompts are suggestions for questions you can ask to encourage students to express their ideas. You may choose to record the ideas for future reference (write on the board and take a photo, write on a transparency or poster, or type on your computer).

After students record their observations on SW 1.1, ask them the following (or similar) questions:

- What did you observe each time you looked in the viewing box? *Accept all observations but make sure some observations include color.*
- What was similar and what was different between your observations? *By making these comparisons students should recognize a pattern: when the green light was shining, green letters were visible; red light, red letters; white light all letters.*

Embedded Assessment in IQWST Curriculum

IQWST materials provide summative assessments as well as numerous opportunities for formative assessments. Many of those opportunities are highlighted in lessons and in the annotated student materials, where questions embedded in the reading, for example, can be used to check understanding. The teacher materials include "educative boxes" (Davis and Krajcik 2005) that provide (1) correct responses, (2) possible answers and what they suggest in terms of student understanding, and (3) ways to discuss a range of responses to clarify student understanding.

The Light Unit

The study described in this chapter concerns the use of the Driving Question Board in the first unit of this comprehensive curriculum, which is a physics unit on light (see Fortus et al. 2006). The driving question for this unit is: "Can I believe my eyes?" The unit content is divided into four learning sets: In the first learning set, composed of five lessons, the major learning goals are to help students to understand how light moves through space, what conditions need to be met for us to be able to see something, and the basic model of the relationship between light and vision. The second learning set is composed of four lessons and deals with the various things that can happen to light when it reaches objects and materials (the interaction of light and matter). The model that was developed in the first learning set is tested, evaluated, and revised in the second one. The third learning set focuses on color, how objects "selectively" reflect and transmit some colors and absorb others, how our eyes perceive mixtures of different colors of light as new colors, and how to determine the color composition of light. At the end of this learning set the students should be able to give a full explanation of the anchoring activity. In the final learning set, the question of what is the physical difference between different colors of light is investigated. Students are introduced to the idea that there are "colors" of light that are not visible (infrared and ultraviolet light). The unit ends by returning to the driving question and discussing the limitations of our eyes and the value of using instrumentation in investigating physical phenomena. The total length of the unit is 6–8 weeks. Figure 8.1 shows the DQB outline as presented in the teacher guide.

Method

Participants

The sixth-grade light unit was piloted in 2005–2006 in eight classes in three different schools in Michigan: one urban (three classes, one teacher), one suburban (three classes, two teachers), and one rural (two classes, one teacher). The majority of the students in the urban and rural schools were from lower to lower-middle income families, of African American and Caucasian ethnicity. In the suburban school most of the students were Caucasian from lower-middle to middle income families. The suburban students were in eighth grade, while the urban and rural students were in sixth grade. Approximately 200 students participated in the pilot.

Figure 8.1. The Light Unit DQB

Instruments and Analysis

All the students completed identical pre- and post-instructional content tests consisting of assortments of multiple choice and open-ended items, sampling a range of cognitive demands. Scoring of tests was done by four raters with inter rater reliability of 0.95.

Interviews were conducted with five urban students, four suburban students, and four rural students about a month after finishing the unit. The teachers were asked to choose students representing different levels, including about the same number of boys and girls.

The interviews, about 30 minutes each, included questions about the main goal of the unit, general content questions, and some attitude questions about their views of the DQB and its purpose. The students were also asked to compare the learning style in this unit to the style used in other units. Student answers to the questions relevant to the DQB were analyzed to find evidence for developing a sense of purpose. The answers were categorized according to the subcategories including making connections, focusing learning, and breaking big questions to small ones. The DQB in each class was photographed, and the questions generated were collected.

Results

Our first research question relates to the way the DQB was used in the different classrooms. Data from observations and videotaped lessons enabled us to compare the process of creating and using the DQB at the three sites.

Creating the DQB: Students' Generation of Questions

In all the classes, students generated questions in the first lesson and sorted them into four categories, which were the main topics of the unit's learning sets. Comparison of questions generated at the three sites shows that students generated similar questions mainly for the first category, the first learning set, on sight and vision (e.g., How do we see? How fast does light travel?), and for the third learning set, on colors (e.g., Do dogs see colors? How does a prism work?). The suburban students generated overall more questions. Also, the focus of each class was different—the rural students had no questions in the fourth learning set category (light we cannot see), and the urban students had fewer questions in the second learning set category (interaction of light with matter). A subject raised in all classes was the connection of light to life and animals, which is dealt with in the IQWST curriculum in sixth-grade biology, seventh-grade physics, and eighth-grade chemistry, but is not a learning goal for this particular unit. The teacher's guide suggests to teachers to keep these questions in a separate "other" category and either use these questions with advanced or interested students as a project, address them at the end of the unit using other resources, or let students know that they will get to these questions in later units (see Figure 8.2).

Different Norms of Using the DQB

While all the teachers referred to the DQB and used it to help students make connections to the driving question and to specific questions and activities, the teachers differed in the way and pace they used this tool. Each teacher interpreted the DQB and the norms of using it in her or his own way, and thus gave it a unique role.

At the urban school the teacher used the DQB mainly to emphasize the connections to the driving question. She used it more often in the first learning set and less in later lessons. At the suburban school the teacher used it as a guide to the unit and also as a tool to keep the class focused. He did not refer to it very often, but used it when moving from one learning-set to another, or to go back to questions and answers. At the rural school the teacher used the DQB almost every lesson, mainly to go back, reference, and review what had been learned. She used to put the answers underneath the questions when they came up. She also attached to it some artifacts that had been created in the class.

We present an example from the urban class, where the teacher presents the DQB and explains how it was constructed. She connects the DQB to the anchoring activity and driving question, and mentions that it is going to be used to focus learning. In addition, she presents norms for using the tool along the unit, telling students they should feel free to come closer and make sure they understand what it means:

Teacher: OK, we're going to move on to our next activity. Before we do that... we're going to focus on how and why our message was hidden, from that you formulated various questions. Our driving question for the unit is, all together, "When can I believe my eyes?" Based on figuring out how that message works [The teacher is showing the class the DQB] how does light help me see? And lastly, is there light I cannot see? All I did was take your questions and post them under

Figure 8.2. Examples of DQB From Different Classes

Urban

Suburban

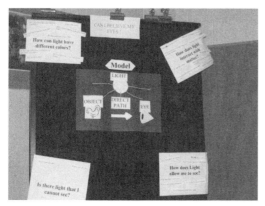

Rural (first year)

Rural (second year)

specific categories, but if you notice you have some questions in blue on white paper, and those are questions we're going to definitely focus on in the unit: How do our eyes let us see? So our question here is how do our eyes let us see? What is the speed of light? And how does light travel?

In our first section we'll focus on these questions, so feel free to come up to the board to look at it and make sure everything is making sense. You're going to look at this board and ask yourself how this question relates to an activity. See if at any point it isn't making sense.

Another use of the DQB by teachers was to review scientific concepts. In the following example from one of the last lessons in the unit, the teacher uses the DQB to review ideas about light in the form of answers to questions written on the DQB:

Teacher:	Let's review what we talked about before the break, as we were studying. What's our driving question?
Student 1:	"Can I believe my eyes?"
Teacher:	Good… If we look at our driving question, when can I (actually) believe my eyes? What are some things we know?
Student 1:	Light speed.
Teacher:	Does anyone remember the number?
Student 1:	200,000
Teacher:	Very close. What else did we cover from that first section?
Student 2:	How does light enable me to see?
Teacher:	Is that what you were reading that question?
Student 2:	We need light to see everything.
Teacher:	How does that work?
Student 3:	The light enters eye goes to retina.

The second research question concerns the influence of the DQB on students' development of a sense of purpose. We were interested to see if the DQB helped students in developing a sense of purpose by making connections, focusing learning, and solving problems. In the following section we present evidence for each of these components.

Using the DQB to Make Connections

There were several types of connections that were highlighted through the DQB:

1. Connecting concepts and activities to the driving question and anchoring activity;

2. Connecting the various activities in the unit to the learning goals;

3. Connecting current topics to previous knowledge, and

4. Connecting concepts and activities to students' daily life.

Connecting Concepts and Activities to the Driving Question and Anchoring Activity

In the following example, the teacher makes a connection between an activity in second learning set (listing the components needed for us to be able to see) and the driving question:

Teacher: Our big question is "How can I believe my eyes?" [pointing to the DQB] and we had a list of things we needed to know or needed to have to see [writing "needed to see" on the board]. What were they?

Student 1:   Eyes.

Student 2:   Light.

Student 3:   Line of sight.

Connecting the Various Activities in the Unit to the Learning Goals

In the following example from the third learning set, a group of students present the model of vision they created and explain how it works:

Teacher: You are going to be listening, you are going to be looking and listening to be certain they show light in constant motion traveling straight from the object into the eye and be certain model shows light into it [reading off overhead].

Student 1:   Our model helps us because it demonstrates how light bounces off object into the eye. Each part is important because if part was missing we wouldn't be able to see the object clearly.

Teacher: What are the parts?

Student 2:   The arrow, straight path, the object represented by car, from the car to the eye light is going into the eye, also there's light bouncing off the car.

Teacher: Great. How well did your group work together?

Student 2:   Our group worked well, everyone in our group had a part in this project by doing different things and helping each other.

Teacher: OK, is there anything you want to add?

Student 3:   We learned that in order to see we need light source, an object and a path.

Teacher: If you wanted to change anything in your model, what would you change?

Student 3:   Nothing.

Teacher: OK, your model is good.

Connecting Current Topics to Previous Knowledge

Teachers used the DQB when teaching new topics to remind students of previous concepts they have learned or questions they have answered before. For example, one of the suburban teachers said during the second learning set:

OK, this is our chart: How does light allow me to see? How do eyes see different depending on the light? Did you mean pupils, or when I looked at the colored message "Can I believe my eyes?" How does animal vision differ from…? These are questions you asked. Does color affect your mood? Ok, why does the light from a camera flash sometimes cause red eye? Ooh, we know that now!

In this example the teacher reminds students about a discussion they had in a previous lesson regarding a question written on the DQB.

Connecting Concepts and Activities to Students' Daily Life

Teacher: OK, if an ink pen is present we can see it, right? How does light itself interact with other objects? What role does light play in allowing us to identify that math book on table two?

Student 4: Shine light to our eyes?

Teacher: So light comes from where? In this case particular source is lightbulb and then it does what?

Student 5: From lightbulb to book and then to our eyes.

Teacher: So that means we have what? In order for light to go from lightbulb to book and then to eyes what does that say about our path?

Student 5: Straight path and nothing else blocking.

Teacher: Anything else we talked about?

Student 6: How light travels.

Teacher: OK, light travels in straight path, how in a straight path? We had toy cars, we made nice lightbulb, how does light travel ahead to the side?

Student 6: In different directions.

Teacher: If we have light here light is going away or towards the source?

Student 7: Away from source.

When the teacher mentions the math book on table two, she contextualizes the concept of light interaction with objects, by indicating a specific object the students can see. When the teacher mentions the toy cars, she refers to a previous activity they did in class, building models to represent light interaction with objects.

There are occasions where students use everyday language or situations in class to make learning meaningful. For example, the following conversation happened during a lesson where students apply the basic model of vision to a new phenomenon—the formation of shadows:

Teacher: Now I have a question for you. Knowing that, and what we did yesterday [the experiment], could we now figure out, if we saw an object, what kind of a shadow it would make? If we were to say yes, how would we decide how big the shadow would be placed in the model? [points to three diagrams drawn on the board, of the experiment apparatus with the object in three positions] I've got three models here: one where it's in the middle, one where it's closer, and one where it's further away. If I were to look at this as a scientist and I had to take the data of the lab I just did, what are some of the things that I could say? What is my evidence?

Student: If you have your hand closer to the light source it would get bigger, and further it would get smaller.

Teacher: OK, let's say "object" instead of "hand."

Using the DQB to Focus Learning

Teachers used the DQB at the beginning of each learning set to focus learning on the main question that is going to be investigated. Another way to use it for this purpose was to highlight or emphasize one quadrant of the DQB. One of the suburban teachers even created separate posters for each quadrant to enlarge the questions she wanted to emphasize.

Teacher: Ok, let's take a look at our DQB. What is our driving question? [students raise hands]

Student: "When can I believe my eyes?"

Teacher: Good, and our focus on this unit is what?

Student: Light.

Teacher: How did you connect that to our opening activity?

Student: We made models.

Teacher: OK, making some models, but ultimately we want to discover about our hidden message, we're going to be focusing on this section the next couple of days. Which question—if you had to pick one which of the four main questions—do you think we're going to be talking about?

Student: "How does light allow me to see?"

Teacher: How can my eyes allow me or let me see?

Using the DQB to Break Questions

Another use of the DQB, which is also one of its main characteristics, was to break big questions to smaller ones. The smaller questions—those that formulate the DQB—are more relevant to student living and can be investigated in classroom activities. The following example is part of the discussion that generated the DQB at the suburban school. The teacher presents the unit learning sets (the "big ideas"), then students raise questions and sort the questions by these categories. The process of generating and categorizing questions has another goal: to scaffold students in learning the skill of asking questions.

Teacher: List your questions. What's our task here? Ask questions about light, color, how we see things…. Take a couple minutes at your table, you and your partner, and come up with three or four questions…. There will be four categories. After we write these questions we'll categorize them. [has students read one of their questions aloud]

Student 1:   How does light help us see different things?

Student 2:   Why did the green apple turn blackish?

Teacher: Why did colored light change the color of the things we were looking at?

Student 3:   How white lights show the real color?

Teacher: Ok, why do you need white light to see the real color? Why do our eyes see things differently in different types of light?

Student 4:   How do you get red eye, what is red eye? How does a flash give red eye?

Student 5:   How do our eyes filter light? Do they filter light?

Student 6:   How does color affect your mood?

Student 7:   Which light can hurt your eyes?

Student 8:   Are animals color blind?

Teacher: Now I want to give you some categories, are you ready for this? put the letter for the category on your paper, you need to think about these and you're going to vote. So A is going to be about vision, seeing, light helping us see [writing on the board], B is going to be how light interacts with materials, C is anything that has to do with color, colored light, and color mixing and D, the last one, is about light we can't see, invisible light.

Students' Content Knowledge

Students' answers to the pre- and posttests were used to check whether they gained content knowledge by engaging in this unit. Paired t-tests of the pre- and posttests of the students in each school show significant growth in students' scores in all the three schools, both for each learning set and for the unit as a whole (see Table 8.1.) The maximum possible score was 44. Overall, the results show significant improvement in all the learning goals at all the schools.

Students' post-interview responses were analyzed to find references to content learned in each of the learning sets. The following are examples of students' responses to the post-interview question *what do you remember learning in this unit with the DQB?* Responses are presented by learning sets:

Table 8.1. Light Unit Pre- and Posttest Scores

	Urban			Suburban			Rural		
	Pre	**Post**	**Effect Size**	**Pre**	**Post**	**Effect Size**	**Pre**	**Post**	**Effect Size**
Total	12.7	23.7	2.6***	16.6	32.6	3.6***	13.7	22.2	2.0***
LS1	6.6	12.1	2.0***	7.6	15.8	2.5***	6.6	9.9	.94***
LS2	5.2	10.1	1.6***	6.5	13.5	3.0***	5.0	9.8	1.5***
LS3	1.4	2.9	1.8***	1.6	4.3	3.4***	1.5	3.1	1.5***
LS4	.5	1.3	1.2***	.8	1.9	1.4***	.6	1.1	.83**

*** p<0.001, ** p<0.005

Learning Set 1 (light and vision):

- How our eyes enable us to see, how they allow us to see.
- Light travels in straight lines, creates a shadow when it cannot pass.
- Light reflects off objects and comes to our eyes.

Learning Set 2 (interaction of light and matter):

- Light can be scattered, transmitted, absorbed, or reflected.
- How light interacts with objects—makes a shadow, goes through transparent objects, or scatters from others.

Learning Set 3 (how we see colors):

- How different colors are shown.
- White light has every color, with a red filter you would get only red. The tree absorbs all the colors except green.

Learning Set 4 (light we cannot see):

- There are UV rays we cannot see.

In general, more students referred to learning set 1 and 2 then to learning sets 3 and 4. In their answers some students mentioned specific questions and answers from the DQB, as well as specific activities they have done in class in order to answer these questions. These results provide evidence for students' ability to connect what they learn to the main ideas, which were presented by the DQB.

Students' Attitudes

Did students find the DQB helpful? In order to answer this question we analyzed interviews with students at the three sites, conducted about a month after they finished the unit. Findings show positive attitudes to the unit in general and the DQB specifically, in spite of the differences between the ways the different teachers enacted the unit and used the DQB.

There was an agreement across students from all schools about the importance of the DQB. They believed it helped them understand better, focus their learning on the main questions, break big questions into smaller ones, organize things without jumping between subjects, making it easy to relate things one to another. The design of the DQB in four sections, and the smaller questions divided by categories helped students understand the purpose of activities and questions in a lesson.

Student answers to the attitude questions were categorized according to the three main characteristics of the driving question board: Making connections, Focusing learning, and Breaking big questions into small questions. In cases where they did not fit any of the categories, answers were categorized as "Other" (see Table 8.2).

Table 8.2. Interview Answers

Category	Student Answers
Make Connections	"It was helpful to keep track of questions we had at the beginning so we knew what we're trying to find out." "It makes it easier to relate back to the question, when all the questions are under one question." "It helped to refer back and remember what we were talking about." "It's kind of hard to understand if you spread out and with the map you connect them together so it's easier." "…concentrate on one section and then do another one and connect them together, so it would be easier for us to understand. I thought it was helpful because it would be hard to understand if they were spread out. Connecting makes it easier to understand. You know what you're focusing on."
Focus Learning	"It helped me understand the questions that we're trying to answer in doing an activity and focus on it. It could help in any unit because it helps you understand what questions you're trying to answer." "The teacher presented it in order to keep students' mind focused on the questions we wanted to learn instead of asking random questions that have nothing to do with the light unit." It helped understand and focus on questions to understand the light unit." "To help us know and organize things more, not jump between sections. It helped me to understand better."
Break Questions	"It was helpful to split the unit to sections we can study individually, so it's easier." "It helped to talk about groups of questions instead of one big question and take one sub-question at a time." "Without the map it would be harder to understand, because it broke questions to smaller ones so we can understand more."
Other	"Without DQB I would probably understand less because it was kind of a guideline of what we're doing." "DQB helped, because in my mind it is easier to understand pictures than words."

Discussion and Conclusions

The DQB is an organizing tool used in the project-based curriculum described in this study, which serves as a visual organizer for all the curriculum's contextualizing features. It was found to foster students' learning of both content and processes. Our results provide strong evidence for improving student content knowledge in all the classes, regardless of social and economic conditions, as well as teacher instruction. Following instruction, students improved their knowledge and were able to connect their learning to the learning goals, as documented in interviews one month after the instruction.

In order to evaluate students' development of sense of purpose we looked at three components, including making connections to the driving question, feeling focused and organized in their learning, and dealing with small questions broken from the big ones. We found evidence in students' interviews for all three components. We found that the use of the DQB supported students' ability to make connections and organize their learning. Most of the interviewed students thought the DQB was helpful for their learning.

The construction of the DQB is an essential activity at the beginning of the unit, but the way it is used is unique to each class and depends on the way the teacher uses it.

In order for learning to be effective it is necessary to combine reform-based curriculum with standard-based teaching, focusing on student understanding and providing scaffolds that fade with time. The DQB was found to be helpful in facilitating this combination. It provides a physical board where individual student interests can be expressed; it assists in focusing on student understanding through the questions they ask; it guides students in connecting the scientific inquiries to the learning goals and developing evidence-based explanations.

Although we focus on the role of the DQB as a contextualizing tool, it is important to keep in mind that we cannot isolate its contribution to developing student understanding. The contributions of other scaffolds, like guidance for discussions that integrate content and practice knowledge, are discussed in other papers (e.g. Weizman et al. 2008, Shwartz et al. 2009).

Overall our results suggest that the DQB was helpful as an inquiry tool and supported student development of a sense of purpose. Further studies are required in order to check the broad and long-term influence of this tool, as well as its use by individual teachers and different subject matter.

The DQB used in this study was a simple board that is available in every classroom. There are advantages and disadvantages to using this type of board compared to more advanced technologies (e.g., computerized boards). For example, it is created by the classroom community and its appearance, as well as its content, is unique to the students of this class and reflects their creativity. However, the idea of a visual organizer can be transferred to different learning environments, and the addition of technological facilities may add other aspects to the role of the DQB.

General findings indicate that students' knowledge in all the classes improved significantly. In addition, the DQB, as a visual organizer of the units' contextualizing features, helped students develop a sense of purpose.

References

American Association for the Advancement of Science (AAAS). 1993. *Benchmarks for science literacy*. New York: Oxford University Press.

Blumenfeld, P., and J. Krajcik. 2006. Project-based learning. In *The Cambridge handbook of the learning sciences*, ed. R. K. Sawyer, 333–354. New York: Cambridge University Press.

Blumenfeld, P., E. Soloway, R, Marx, J. S. Krajcik, M. Guzdial, and A. Palincsar. 1991. Motivating project-based learning: Sustaining the doing, supporting the learning. *Educational Psychologist* 26 (3/4): 369–398.

Davis, E., and J. Krajcik. 2005. Designing educative curriculum materials to promote teacher learning. *Educational Researcher* 34 (3): 3–14.

Deci, E. L., and R. M. Ryan. 1987. The support of autonomy and the control of behavior. *Journal of Personality and Social Psychology* 53: 1024–1037.

Edelson, D. C., D. N. Gordin, and R. D. Pea. 1999. Addressing the challenges of inquiry based learning through technology and curriculum design. *Journal of the Learning Sciences* 8: 391–450.

Fortus, D., B. Hug, J. S. Krajcik, K. Kuhn, K. L. McNeill, B. Reiser, et al. 2006. Sequencing and supporting complex scientific inquiry practices in instructional materials for middle school students. Paper presented at the National Association for Research in Science Teaching conference, San Francisco.

Harris, J. C. 2005. Investigating teaching practices and student learning during the enactment of an inquiry-based chemistry unit. Unpublished doctoral dissertation. Ann Arbor, MI: University of Michigan.

Kozma, R. B. 1991. Learning with media. *Review of Educational Research* 61: 179–212.

Krajcik, J. S., P. C. Blumenfeld, R. W. Marx, K. M. Bass, J. Fredericks, and E. Soloway. 1998. Inquiry in project-based science classrooms: Initial attempts by middle school students. *Journal of the Learning Sciences* 7: 313–350.

Krajcik, J., C. Czerniak, and C. Berger. 2002. Teaching science in elementary and middle school classrooms: A project-based approach, 2nd ed. Boston, MA: McGraw-Hill.

Krajcik, J., K. L. McNeill, and B. J. Reiser. 2008. Learning-goals-driven design model: Developing curriculum materials that align with national standards and incorporate project based pedagogy. *Science Education* 92: 1–32.

Linn, M. C. 2000. Designing the knowledge integration environment. *International Journal of Science Education* 22 (8): 781–796.

Marx, R. W., P. C. Blumenfeld, J. Krajcik, and E. Soloway. 1997. Enacting project based science. *The Elementary School Journal* 97 (4): 341–358.

National Research Council (NRC). 1996. *National science education standards*. Washington, DC: National Academies Press.

Rivet, A. E. 2003. Contextualizing instruction and student learning in middle school project-based science classrooms. Unpublished doctoral dissertation. Ann Arbor, MI: University of Michigan.

Shwartz Y., A. Weizman, D. Fortus, J. Krajcik, and B. Reiser. 2008. The IQWST experience: Using coherence as a design principle for a middle school science curriculum. *Elementary School Journal* 109 (2): 199–219.

Shwartz, Y., A. Weizman, and D. Fortus. 2009. Talking science, classroom discussions and their role in inquiry-based learning environments. *The Science Teacher* 76 (5): 44–47.

Singer, J., R. W. Marx, and J. S. Krajcik. 2000. Constructing extended inquiry projects: Curriculum materials for science education reform. *Educational Psychologist* 35 (3): 165–178.

Weizman, A., Y. Shwartz, D. Fortus and J. Krajcik. 2008. Developing the practice of scientific modeling through classroom discussions. Paper presented at the National Association for Research in Science Teaching conference, Baltimore.

Wenger, E. 1998. *Communities of practice: Learning, meaning, and identity*. New York: Cambridge University Press.

Wiggins, G. P., and J. McTighe. 1998. *Understanding by design*. Alexandria, VA: Association for Supervision and Curriculum Development.

Communic–Able:

Writing to Learn About Emerging Diseases

Andrew J. Petto
University of Wisconsin–Milwaukee

Setting

High school students in a seven-week writing-intensive project at the University of Wisconsin School of Medicine and Public Health serve as research apprentices. This project focuses on students in need of additional development in communication, computation, or scientific background, and creates a research group that carries out original research on a mock epidemic. To succeed in the program, students must master at least four—but up to seven—forms of scientific communication, including news releases, public information brochures, weekly newsletters, project web pages, and formal written and oral research presentations. The challenge of clear communication of scientific information to a variety of audiences and in several formats tests the students' understanding of the research foundations and methods of inquiry appropriate to their studies and the health research field in which it is grounded. Success of this project is outcome-based, and the students in this "developmental" group perform on par with other students in the program.

Introduction

The Research Apprentice Program (RAP) is a summer research experience for students in grades 10 and 11 at the University of Wisconsin School of Medicine and Public Health (SMPH) in Madison. The program is based in the Office of Minority Affairs in the SMPH Office of Admissions and actively recruits students from underrepresented populations throughout southeastern Wisconsin. Participants typically participate individually or in pairs in active biomedical research projects with UW researchers. RAP also provides support for these students in daily sessions on scientific writing, mathematics, library research, and computer skills.

Beginning in 2003, RAP added a new project originally conceived as a way to include applicants who met most, but not all, of the program requirements. Based on their application materials, high school transcripts, and letters of recommendation, these students are judged to be highly motivated and strongly interested in biomedical research, but in need of additional development in their communication, computation, or scientific preparation. These students would work as a team with a research mentor in a program that would focus on basic academic skills while guiding the students through the research process. Because RAP requires all its participants to prepare a formal research paper and an oral presentation at a mock scientific meeting, this new project emphasized the skills required for effective research communication: exploring the scientific literature, critical reading, preparation of research outlines, and writing for difference audiences and in different formats. Much of the approach follows closely that described in Penrose and Katz (2004). The project described here grew out of the initial experience with the mock Tickettsia outbreak in the summer of 2003 and began in 2004.

This project uses a problem-based learning (PBL) approach (Chin and Chia 2008; Massa 2008), applying research and writing skills to the problem of the emergence of a mock communicable disease on the UW campus. Students in this project form an epidemiological task force called U-WHO (University of Wisconsin Health Organization) and design their own projects to study the spread of this imaginary disease, based on research approaches that they have explored in the scientific literature. Since its inception in 2004, only a few students have chosen a project identical to those in previous years (See Table 9.1). In cases in which the research problem is similar, we treat the previous years' papers as a part of the research literature.

For this project, each student participates in multiple writing activities, learning to communicate about the emerging disease and disease-control efforts in formal and informal environments modeled on the communications about infectious diseases at the U.S. Centers for Disease Control and prevention (CDC; *ww.cdc.gov*). These components of public health communication for the CDC and for RAP are shown in Table 9.2, page 134. Like the CDC model, successful communication in this project requires an awareness of the audiences for the writing, and we follow the authentic assessment strategy of Gunel et al. (2009) in assessing students' success.

The assessment of the students' learning is based on the concept that clear, effective communication provides evidence of successful learning. Students who write accurately about the nature of emerging infectious diseases and strategies for their control and eradication can only do so if they have mastered the underlying scientific knowledge that is the basis of their research (Gunel et al. 2009). In essence, this approach follows the common expression regarding writing across the curriculum courses: "Good writing is good thinking" (Oakley 1999; see also comments by Uggen and Hartman at *http://contexts.org/articles/summer-2008/good-writing-and-good-thinking*).

This component of the RAP program has the dual "rhetorical" and "content" learning goals associated with most "writing-to-learn" programs (Gunel et al. 2009). This discussion will focus on students' abilities to demonstrate success in the rhetorical: the structures, forms, and conventions of effective scientific communication for different audiences who have—or ought to have—an interest in the outcome of the project that the students are pursuing.

Methodology

After learning about emerging infectious diseases and their control, RAP students formulate their mock virus for the epidemic. Based on a suite of characteristics of infectious diseases, students decide how it is transmitted (directly or indirectly), whether immunity and cross-immunity are possible, whether infection persists or recovery is possible, and so on. After these decisions are made, team members create "virus" particles from a roll of numbered tickets, and place a brief summary of the instructions on the back (see Figure 9.1, p. 134).

Each team member receives a unique range of numbers from the roll of tickets. These ranges correspond to subtypes or strains of the virus. In some years, team members have also

Table 9.1. RAP Emerging Disease Research Topics[1]

2004	2005	2006	2007	2008	2009
Geographic Patterns of Infection Rates	Ethnic Differences in Infection Rates and Reporting	Effects of Incentives on Reporting Infection	Risk of Infection by Type of Physical Contact	Vaccine Development	Geographic Models of Disease Spread
Comparing Effectiveness of Public Communication Methods	Prevention and Control of the Spread of New Infectious Diseases	Sociocultural Aspects of Infectious Disease Reporting	Transportation Systems as Routes for Disease Transmission	Effectiveness of Public Health Communications	Behavioral Change in Response to Public Health Communication
Effects of Disease Awareness on Risky Behaviors	Potential for Vaccine Development	Comparing Active and Passive Acquisition of Infection	Behavioral Differences and Risk of Infection by Sex	Disease Risk in New vs. Established Social Relationships	Internal vs. External Spread of Disease in Interconnects Social Networks
Differences in Risk Between Commuting and Residential Students	Bioterrorism Potential of Emerging Infectious Diseases	Levels of Herd Immunity and Spread of Disease	Comparing Risk of Infection in Work vs. Social Environments	Sex Bias in Disease Transmission	Sex Bias in Disease Transmission
Effects of Information Outreach on Risky Behavior	Adolescent Social Networks and Disease Spread	Differential Virulence of Subtypes	Differential Virulence of Subtypes	Differential Virulence of Subtypes	Social Network Participation and Sociability-Related Risks of Infection
	Age-Related Risks in Disease Transmission	Effects of Weather-Related Behaviors on Risk of Infection		Intimacy of Social Relationships and Risk of Infection	Effects of Weather-Related Behaviors on Risk of Infection
	Ethnicity as the Basis of Social Networks for Disease Transmission	Global Monitoring to Identify Risks of Disease Spread		Geographic Patterns of Infection Rates	

The total number of students in this project varied from year to year.

Table 9.2. RAP Public Communications Activities

CDC	RAP
Website	Website
Mortality and Morbidity Weekly Report	Weekly Update
Disease Information Brochures	Disease Information Brochures
Disease Information Posters	Disease Information Posters
Public Information Meeting	Public Information Meetings
Formal Research Reports (original papers and presentations)	Formal Research Reports (original papers and presentations).
Links to Local and Global Resources	Links to Local and Global Resources
News Briefs and Summaries of Current Research	Précis of Team's Current Research (individual summaries)
Disease-Specific Control and Prevention Information	Control Information for Selected Communicable Diseases

Figure 9.1. The YPK virus from 2008
(Each year, students give the new virus a different name.)

Figure 9.1a. The front of the "virus" particle.

Figure 9.1b. The back of the "virus" particle with instructions about its characteristics.

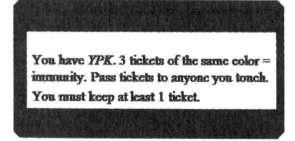

added a small colored dot to make it easy to distinguish among subtypes when collecting data about the spread of infection. Once the tickets are ready, students prepare public information brochures, posters, and fliers, which they distribute primarily to other students in special summer programs on campus (a sample brochure from a past program is included in Appendix A). Then the team members begin to distribute the tickets to start the epidemic.

At the end of each week, team members conduct a disease surveillance activity, usually at meal times. At these activities, team members ask summer program participants if they have received any tickets and whether they would be willing to have the team members record the ticket numbers in their possession. Those who agree to be counted in the surveillance receive a coupon that can be exchanged for a small ice cream cone at the store on the UW campus.

After each surveillance session, the team's data captain coordinates data entry and a summary of the cases observed to date. These data are used to generate a weekly report that is distributed to all the at-risk populations. The data also provide the basis for the individual research projects that the team members will carry out over the course of the summer.

Results

Students begin with basic readings in epidemiology (CDC 2004; Coggin, Rose, and Barker 1997) and background on how to read and interpret a scientific research paper (MacNeal 2001). Their first task is to find one type of epidemiological study that they would like to perform, and, with the help of instructors, they begin to locate background information that they can use to design and carry out their own studies.

Students begin writing and working with data immediately. Their first assignment is to learn to write a complete profile of an infectious disease and strategies for its control by following the format in the *Control of Communicable Diseases Manual* (Heymann 2008). These profiles are based on diseases of interest to individual students. Each student becomes the team's "expert" regarding a particular communicable disease. Once the profiles are complete, they are posted to the team's website (the 2009 website can be viewed at *https://pantherfile.uwm.edu/ajpetto/www/ RAP_2009/RAP_2009_Home.htm*).

There are several concurrent writing tasks in this program. The brochure (Appendix A) and the website are designed collaboratively, but there is also one student who is primarily responsible for design and production. As they prepare their own research projects, team members begin with writing a brief problem statement that appears on the project's website (*https://pantherfile.uwm.edu/ajpetto/ www/RAP_2009/RAP_2009_Home_files/Page348.htm*).

During the remaining weeks of the program, team members engage in individual and collaborative writing, preparing weekly summaries of the spread of the epidemic and the characteristics of those reporting to the surveillance sessions. Each weekly summary requires analysis of new data and revision of the earlier summaries based on the current sample. All the summaries are printed and distributed throughout the campus, and

RAP students collaborate to review recent surveillance data.

they are also posted on the team's website (*https://pantherfile.uwm.edu/ajpetto/www/RAP_2009/ RAP_2009_Home_files/Page947.htm*), which is updated each week.

Students also begin the process of constructing their formal scientific writing with the preparation of a research abstract, a brief bibliography of primary sources, and the introduction section of their formal papers (including the review of literature and problem definition). Once they have collected sufficient data, students learn to use basic statistical analyses (Daniel 2007; Milton 1999) to evaluate the outcomes of their projects. These become the basis of their Methods and

Results sections. Once these sections are revised and approved, students move on to finish the Discussion and Conclusions sections.

At each step, students also practice extracting key points for use in their final formal oral presentations, assemble visual materials, and develop charts and graphs to support their presentations. Some of these materials are derived from the materials provided in the weekly reports, but others are generated especially for the oral presentation, which is organized as a contributed paper session at a professional meeting.

This project emphasizes critical reading and writing. Students begin with an area of interest that reflects public health research methods and practices that engage them individually. Once the students have identified the sort of study they wish to do, the instructor works with students and library staff to identify relevant resources. Then, based on a few key studies that they have read, students frame their own studies in terms of the *Tickettsia* outbreak on campus and write a preliminary abstract laying out the main topic area, the key variables and methods of collecting and analyzing them, and the contribution they hope this study will make to the control of the mock epidemic. After instructor's feedback, students post these abstracts on the project website (*https://pantherfile.uwm.edu/ajpetto/www/RAP_2009/RAP_2009_Home_files/Page348.htm*).

One of the most important transitions for students is to move from reading and reporting on the research literature that they have read to *applying* the key elements of the research literature. For example, a number of the articles they read at the outset are review articles that describe a variety of methods for collecting and analyzing data. Only some of these are likely to be relevant to the specific studies that the students will carry out, so the goal is to have students learn how to use review papers, methodology papers, and meta-analytic papers as resources that will help them succeed in their own projects.

An important part of this process is learning to extract key elements from the research literature and note its applicability for the students' own work. To assist in this process, students use two different forms for tracking the literature that they read. The first form is a literature review summary that collects important information from individual papers (an example of this form is available at *https://pantherfile.uwm.edu/ajpetto/www/Lit_rev.htm*). The most important part of this form for students is that it requires that they compare the study they are reading with other studies and note whether (and how) it differs in conclusions, methods, or any other significant features. Successful use of this form helps students to see that the research literature contains a great deal of variety and that the conclusions in any particular study may provide only a part of the solution to an important problem.

The second part of the students building their research literature background is to identify the "life cycle" of research issues. It is often the case that the predominant issue in a particular field changes as more research is completed or as new techniques, models, understandings, and results are available (Tracking Trends in Literature form: *https://pantherfile.uwm.edu/ajpetto/www/trends.htm*). Students complete this form in a word processing program or spreadsheets, then sort the form by the publication date of the article in their citations. When an issue is scientifically unresolved, it will be prominent in the table, but as its resolution is presented and accepted, it will cease to appear very often. These trends help students to understand which models, hypotheses, and analytical practices are currently in use.

These steps are completed as students develop their research interests and begin to design their studies. They learn what types of conclusions can be drawn from different types of data and begin to refine their studies as they write their results. They build a research plan, collect and analyze data, summarize their findings in tables, charts, and graphs, and begin writing about their epidemiological specialties in the first week of the program. In the end, they create a standard research paper and prepare to make an oral presentation that is a part of a miniature scientific research meeting.

The research process that the students follow is iterative. As they begin to write about their research problems, they learn to identify the areas that need more data, more analysis, or perhaps more development. Rather than the linear model of *the* scientific method for research, these students learn that scientific research is best described as a web or a network with multiple pathways to the final outcome—and the need to retrace some of these pathways from time to time. As an example, we use The *Real* Process of Science flowchart found at the University of California, Berkeley website: *http://undsci.berkeley.edu/article/0_0_0/howscienceworks_02.*

Evidence of Success

We take the approach that writing is an integral part of the *process* of scientific inquiry and that writing itself is an important process. Therefore, the goal for the students is to learn the elements of effective communication as they apply them to the different types of writing that they must do for the program. These different outcomes are shown in the right column of Table 9.2, page 134.

The rubric used for assessing student writing is multidimensional and iterative. The instructor responds to student writing by noting areas that need additional work and areas that have been completed successfully (see Table 9.3). Students then focus on the areas that need additional work. At the outset, the assessment focuses on major issues, such as content, organization, format, and audience at the outset (including proper scientific vocabulary). Once these are resolved, the focus shifts to issues of presentation of graphical or tabular features and the documentation of citations and references. Finally, the focus shifts to issues of clarity, the technical aspects of punctuation, grammar, mood and voice, and word economy.

Table 9.3. Sample Formative Rubric for Assessing Student Writing

Focus Correction Area(s)	Identified by Instructor	Action by Student	Review by Instructor	Comments
Content				
Structure				
Paragraphs				
Charts/Tables				
Citation				
Mood, Voice				
Economy				
Vocabulary				
Punctuation				

In practice, however, the instructor focuses with the student on the particular issues that are the most pressing in the individual pieces of writing. It is much more like the "real" process of science illustrated in the UC–Berkeley flowchart: iterative, interactive, and nonlinear. The initial response from the instructor is to point out a few items that need correction, and then make suggestions for the type of revision that is most appropriate. Students respond often by finding their own ways to address the issue—without always applying the specific suggestions of the instructor. This is perhaps the most valuable outcome—since students craft their own solutions to improving their writing rather than just copying what the instructor suggested—and is what Klein (2004) considers the main evidence of the transformation of learning through the process of writing. Examples of student writing are in Appendix B, page 142.

In the first example, we see a focus on improving clarity and word economy in the introduction to the project. The goal is for the student to identify the larger context in which the study is being conducted and then introduce the reader to what the student's project might contribute to our understanding. The first revision of the student's draft shows a clear identification of the main problem (risk of disease spread through social contact) and which aspect of social contact will be examined in the study (social networks). The next paragraph begins the process of literature review, based on a few key studies that apply social networking models to epidemiology.

The second example looks more closely at the literature review. After examining several research articles, the student needs to tell the reader how these results relate to the study that he or she has completed. In the revision, the link between the research literature and the variables that the student will study are made explicit. Later, the student's methods section shows how these variables will be observed, measured, and analyzed. Furthermore, this very clear statement of expectations from previous research literature will allow the student to assess whether this study met those expectations and to what extent the results of this study match those reported in the literature.

The third example shows a student learning to use the research literature more effectively. The goal is for students to realize what aspects of previous research are most relevant for their own studies. In this case, the original draft dwelt too much on details of previous research that did not relate directly to the student's project. In the final revision, it is much clearer that the aspect of the research literature to which the student is referring is the effect of public information campaigns on awareness and health-related behaviors of those who are exposed to them.

The final example is from the discussion section of a student paper. This study was similar to a study done in a previous year, so the discussion includes a comparison of the results from both years. The students are expected to review the results and then to compare them with expectations and with the findings in the research literature. The second revision shows more clarity in both of these areas after focusing on the summary of the results and their alignment with previous studies and with expectations from this study. In the next iteration, the student and instructor worked on the final paragraph to clarify the significance of directional contact related to sex ratios and the comparison with the prior year's findings.

In addition to the instructor's comments and suggestions, students also have "outside" readers: staff members within the program who provide support for the research projects and process for all the students in the programs. They respond to the students' writing both as instructional staff and as a disinterested general audience. Their reactions to the student writing are a gauge of the students' success in this part of the program.

Conclusions

The U-WHO project in the RAP program incorporates a problem-based learning approach to the biomedical research process. Students in the U-WHO project participate in an intensive research and writing program focused on the prevention and control of the spread of a mock virus on the campus of the University of Wisconsin, Madison, learning about the process of scientific research while they are

RAP student makes final oral presentation of her research project.

learning to communicate effectively to a variety of audiences, including other researchers, the general public, news outlets, and public health professionals. This ungraded program relies on the students' own written work, guided by developmental rubrics and based on the professional conventions of writing in relevant fields. The students' successful outcomes are evident from the projects' websites as well as their formal research presentations—their research papers and their oral "conference" presentations.

The students chosen for this project are generally judged to need extra help in communication, computation, or basic scientific knowledge, but otherwise showing good motivation and interest in biomedical sciences. Improvements in student communication are evident when comparing early and late drafts of the students' formal papers. Furthermore, the final research papers and oral presentations produced by the students in this "developmental" group are on par with those produced by the other students in the program.

To succeed in this program, each student must demonstrate successful written communication in different writing formats and styles—all of which relate to NSES Goal 3: "Engage intelligently in public discourse and debate about matters of scientific and technological concern" (NRC 1996, p. 13). In the process, successful students learn both effective communication skills *and* the process of scientific inquiry that informs intelligent discourse and cogent persuasion in matters of scientific and technical concern.

Acknowledgments

We are grateful to Joan Brooks, who originally conceived of this new project and recruited us to participate in and to develop this extension of the RAP experience. This project could also not have been successful without the encouragement and feedback in the early years from Elizabeth Haslam in the School of Education at Drexel University and the advice and support of Lili Velez and Kelleen Flaherty, who were my colleagues in the Biomedical Writing Program at the University of the Sciences in Philadelphia during the time that this special project was in its beginning stages. I am grateful to Beth Goodbee of the UW Writing Center for valuable comments and suggestions on drafts of this chapter. Finally, we thank Gloria Hawkins, assistant dean in the department of academic affairs at the UW Medical School and director of RAP for the opportunity to develop and carry out this project.

References

Centers for Disease Control (CDC). 2004. What is epidemiology? *http://www.cdc.gov/excite/PDF/intro_epi.pdf.*

Coggin, D., G. Rose, and D. J. P. Barker. 1997. *Epidemiology for the uninitiated.* 4th ed. London: BMJ Publishing. *www.bmj.com/epidem/epid.html*

Daniel, W. W. 2005. *Biostatistics: A foundation for analysis in the health sciences.* 8th ed. New York: Wiley.

Gunel, M., B. Hand, and M. A. McDermott. 2009. Writing for different audiences: Effects on high-school students' conceptual understanding of biology. *Learning and Instruction* 19 (4): 354–367.

Heymann, E. L., ed. 2008. *Control of communicable diseases manual.* 19th ed. Washington, DC: American Public Health Association Press.

Klein, P. 2004. Constructing scientific explanations through writing. *Instructional Science* 32: 191–213.

MacNeal, A. 2001. How to read a scientific research paper: A four-step guide for college students and for faculty. *http://hampshire.edu/~apmNS/design/RESOURCES/HOW_READ.html*

Milton, J. S. 1999. *Statistical Methods in the Biological and Health Sciences.* 3rd ed. Madison, WI: McGraw-Hill.

National Research Council (NRC). 1996. *National Science Education Standards.* Washington, DC: National Academies Press.

Oakley, T. 1999. Copious reasoning: The student writer as an astute observer of language. In *Language Alive in the Classroom*, ed. R. S. Wheeler, 130–138. Abington, UK: Praeger.

Penrose, A. M., and S. B. Katz. 2004. *Writing in the sciences: Exploring conventions of scientific discourse.* 2nd ed. New York: Longman.

Appendix A—Example of Student-Written *Rapcitity* Public Information Brochure

WHAT

This is an asymptomatic disease in which the infected students don't feel any symptoms. Students who receive or find tickets will have this disease permanently. Anyone with tickets can transmit RAPCITITY to others. (See method of transmission under HOW.)

MORE INFO

For more information contact:

I.A.

A.A.

J.F.

D.G.

E.G.

K.G.

M.K.

P.R.

WANTED

RAPCITITY
ON THE
LOOSE

WHO

As a field unit of the RAP CDC (Center for Disease Control), we are currently investigating a RAPCITITY cluster in the UW-Madison campus. We are aware that the disease outbreak began in Slichter Hall among High School students in the following programs: RAP, NASA, SSI, ESP.

HELP!

We need your help to track the spread of the disease and control it. We will be setting up booths in the Carson building cafeteria every Tuesdays and Thursdays to record your ticket number. Your participation will not go unrewarded.

HOW

METHOD OF TRANSMISSION

Please follow these rules when you transmit the disease.

- Anyone with at least one ticket has immunity (and can't receive more tickets).
- If you have physical contact with someone without tickets, give him/her half of your tickets.
- If you have contact with an object that others may have contact with, leave behind two tickets.
- If you have one ticket left, you can't give it away.

Appendix B—Examples of Student Writing Outcomes

1. From the introduction to a paper on the study of social networks in spreading disease.

Original Draft

The main problem is controlling the spread of Oompapikitis, a communicable disease. Oompapikitis is a virus that is spreading all over campus. It is a virus that was made to display virtually how disease spreads from person to person. The study will be on how the disease spreads based on social networks and learning how to control it. Communicable diseases spread based on social contacts. Social networks can modify and direct the spread of these diseases, so identifying social networks can help to estimate risk of infection and provide a strategy for controlling epidemics. Because the disease Oompapikitis is spread based on direct physical contact, it will be a good example of how social networks are used in transmitting disease.

Many previous studies have been done on the idea of social networks. There are many types of social networks that are used in obtaining research. One idea that will be used in this study is the idea of small-world networks (Friedman and Aral 2001). Small-world networks are essentially, networks that are connected by one or two key people who have a significant hold in both of the networks or groups that they are in. It confirms the thought that not only do people have ties within their own social network; but that they also have weaker ties outside of their networks those they still are in contact with (Luke and Harris 2007). This will probably help in the study since it is almost guaranteed that two different summer programs may contain two people who are friends in each.

First Revision

Communicable diseases can be spread based on social contacts. Social networks can modify and direct the spread of these diseases, so identifying social networks can help to estimate risk of infection and provide a strategy for controlling epidemics (citation here). On the UW campus, Oompapikitis is a mock epidemic that began in the summer of 2009. Because the Oompapikitis is spread by direct physical contact, it will be a good candidate for learning how social networks function in transmitting disease. This study will examine how the disease spreads based on social networks and how this information can help us learn how to control it.

Many previous studies have been done on the idea of social networks, and there are several models of social networks that can be applied in epidemiology research. One model that is small-world networks (Friedman and Aral 2001). Small-world networks are essentially two or more social networks that are connected by one or two key people who have a significant hold in both of the networks or groups that they are in (Luke and Harris 2007).

2. From a discussion of relevant research literature.

Original Draft

In studies conducted in gender by Hall (1984), she suggests that females tend to be more inclined to initiate touch, and they also are generally more likely to be the individuals that receive the touch. Similar results are present in same-sex pairs; the female pairs experienced a higher frequency of touches than the male pairs. In an additional study done by Henley (1977), he determined that the results are seen because women usually respond more positively to touch than males in most situations.

.

First Revision

Hall (1984) suggests that females tend to be more inclined to initiate touch, and they also are generally more likely to be the individuals that receive the same sex and mixed sex touches. Similar results are present in same-sex pairs; the female pairs experienced a higher frequency of touches than the male pairs. Henley (1977) suggests that these results are seen because women usually respond more positively to touch in initiating and receiving than males in most situations. Based on these studies, I expect that more females than males will be involved in spreading disease. I expect to observe more female-initiated touching in the locations where the most infections have been reported so far.

3. From a discussion of relevant research literature.

Original Draft

Keating and Alfred Adewuyi did a study on awareness of HIV/AIDS, based on exposure to media reports, would make a significant difference in the risk of getting the virus and on the rate of passing the virus to another.

In response to the growing HIV epidemic in Nigeria, the U.S. Agency for International Development (USAID) initiated the VISION Project, which aimed to increase use of family planning, child survival, and HIV/AIDS services (citation here). The VISION Project used a mass-media campaign that focused on reproductive health and HIV/AIDS prevention. to the main goal was to investigate the effect of program exposure on 1) discussion of HIV/AIDS with a partner, 2) awareness that consistent condom use reduces HIV risk, and 3) condom use at last intercourse. Exposure to the VISION mass media campaign was high: 59%, 47%, and 24% were exposed to at least 1 VISION radio, printed advertisement, or TV program about reproductive health, respectively. Those with high program exposure were almost one and a half times more likely than those with no exposure to have discussed HIV/AIDS with a partner. However, the study concluded that the greatest effect would be to combine mass media campaigns "done in conjunction with other interventions, and targeted towards individuals with specific socio-demographic characteristics."

Final Revision

Keating and others (2006) studied whether awareness of HIV/AIDS, based on exposure to media reports, would make a significant difference in the risk of getting the virus and on the rate of people passing the virus to one another. In their study of the U.S. Agency for International Development, VISION Project, they noted that those with highest exposure to one or more media messages were almost one and a half times more likely than those with no exposure to have discussed HIV/AIDS with a partner. The study concluded that the greatest effect would be to combine mass media campaigns with other interventions targeted towards individuals with specific socio-demographic characteristics." (Keating et al. 2006)

Gonzales and others (2008) also used mass media as a way to inform people about certain different types of diseases. The challenge was to inform the public about antibiotic resistance in community acquired bacterial pathogens (e.g., *Streptococcus pneumonia, Staphylococcus aureus, and Escherichia coli*). The main goal was to reduce the inappropriate use of antibiotics.

This and a number of other controlled studies in the U.S. (reported in Gonzales et al. 2008), confirmed the benefit of combined patient and physician educational interventions in reducing antibiotic use within a variety of health care delivery systems. (Gonzales et al. 2008)

4. From the discussion section of the formal research paper.

Original Draft

With in this study, females have been proven to be the critical group in transmission of Oompavirus in sex-based networks. Males make up almost half the susceptible population they only reported about 33% of infections in 2009 (and only about 11% of infections in 2008). This significant change in sex rations allowed me to conduct a more reliable study. The chi-square exemplified that the patterns are significantly similar in both the previous and present results.

Based on these results FtF (female-to-female) transmissions are the core of the Oompapikitis outbreak; this is consistent with their having higher risk behavior— which in this case is physical contact. Females are found to be closer or more imitate in close relationships which could result in the high increase in the infections. These results are similar to sex differences in risk profiles reported in the literature (for example, see Bettinger, Adler, Curriero, and Ellen 2004). Giordano (2003) demonstrated that sex is a specific variable in the relationships of the networks and affects both the intensity and the direction of interactions.

On the other hand, the transmission of MtM was higher than seen previously. The previous study's MtF transmission in 2008 was the transmission that was also higher than expected. In contrast to our expectations, this data showed that females are most at risk from same-sex, but not in opposite sex transmission which is similar from last year. From last year's analysis I expected females to be more involved in spreading the epidemic, but I did not expect that they would not also be at the highest risk of becoming infected. This was very shocking because I was expecting the MtM transmission to be less based on the fact that there were more females, although the ratio of females to males was not significantly different.

Second Revision

In this study, females are at highest risk of infection from transmission in sex-based networks. Males make up almost half the susceptible population, but they only reported about 33% of infections; last year it was only about 11% of infections in the entire infected program population. This shows great increase within the male infected population and allows me to conduct a more valid experiment. The chi-square analyses showed that the role of females in transmission of the infection was similar in both the previous and current years. Based on these results FtF (Female-to-Female) transmissions are the core of the Oompapikitis outbreak; this is consistent with their having higher risk behavior — which in this case is physical contact. These results are similar to sex differences in risk profiles reported in the literature (for example, see Bettinger, Adler, Curriero and Ellen 2004). Giordano (2003) demonstrated that sex is a specific variable in the relationships of the networks and affects both the intensity and the direction of interactions. Girls are found to be closer or more imitate in close relationships which could result in the high increase in the infections.

On the other hand, the transmission of MtF (male-to-female) was higher than anticipated. The previous study's MtF transmission was also higher than expected. In contrast to our expectations, this data showed that females are most at risk from same-sex, but not in opposite sex transmission which is similar to results from last year. From last year's analysis I expected females to be more involved in spreading the epidemic, but I did not expect that they would not also be at the highest risk of becoming infected. This was very surprising because I was expecting the percentage of MtM (male-to-male) transmission to be lower based on the fact that there were more females, although the ratio of females to males was not significantly different.

Using Socioscientific Issues as Contexts for Teaching Concepts and Content

Scott Applebaum
Palm Harbor University High School

Dana L. Zeidler
Robinson High School

Kristy Loman Chiodo
University of South Florida

Setting

This chapter focuses upon the conceptual development and implementation of a socioscientific issues (SSI) curriculum in two high school science classrooms. The scenarios and perspectives described in this chapter exemplify fundamental examples of best practices in the SSI tradition. Teachers enter the science education program at the University of South Florida for several reasons, including satisfaction of certification requirements and job performance improvement. As expected, students in master and doctoral programs enter the science education programs with differing personal views about education in general, and science teaching in particular. A look at these differences and their effects on teaching will be explored.

Introduction

The University of South Florida has developed a research program that uses SSI as context for teaching concepts and content; specifically, themes from different disciplines that have explanatory power related to how children think, reason, and learn. The program is designed for graduate students to discover the importance of building a theoretical framework that includes concepts from philosophy, sociology, psychology and education. The concepts are fitted into a framework that is cohesive, internally consistent, and engaging for children. Using socioscientific issues creates an interactive learning experience, engaging students in sustained dialogue, discussion, and debate. The topics are typically controversial in nature, requiring a degree of moral reasoning and the evaluation of ethical and practical concerns. Above all, the intention is for science issues to become personal, meaningful, and engaging to students, while requiring the use of evidence-based reasoning and contemporary context for understanding scientific information (Zeidler 2003; Zeidler and Sadler 2008; Zeidler and Nichols 2009). Within this framework, students are better able to match their actions with high teacher expectations in a classroom community that entails the joint construction of social and scientific knowledge. Conceptual understanding, reflective reasoning, and understanding of the nature of science are enhanced. Students are afforded more responsibility for their own learning, becoming more persistent and autonomous—consistent with developing a sense of virtue (Fowler, Zeidler and Sadler 2008; Walker and Zeidler 2007; Zeidler and Sadler 2008; Zeidler et al. 2009).

Students enrolled in the science methods class engage in case studies, controversial topics, and nature of science (all part of the SSI framework). Preservice and inservice teachers taking a Contemporary Issues and Trends course learn about the theory base for SSI, and use the research literature to develop full-fledged SSI activities for their grade levels and subject matter. Doctoral students become immersed in the research base that connects to SSI, including course work in epistemology, the philosophy and nature of science, moral and ethical issues in science education, and reasoning and cognition. Whether in methods courses or graduate seminars, there is an array of reactions to what students perceive as challenges to their comfort zones about what constitutes contemporary science teaching. Reactions include indifference, skepticism, openness, and enthusiasm. Many recognize the paradigm shift from a teacher-centered to a student-centered inquiry classroom (see Zeidler 2002). As a result, students identify the research and philosophies driving contemporary teacher education and use evidence-based arguments to develop pedagogically sound methods of SSI instruction.

Teacher Views: Understanding the Socioscientific Issues Curriculum

The authors entered the field of education from backgrounds and professions that required expert scientific knowledge; however, our understanding of education, teaching, and curriculum was limited to our experiences as students. Independently, we sought to acquire the skills that good teachers possess, acutely aware that there was no singular vision or method of teaching science. We quickly discovered that teaching is multifaceted and untidy; teaching styles are inevitably individualized, personally motivated, and focused upon the cognitive, emotional, and social needs of the students. Typical of first-year teachers, we believed that teaching science content is a predominant role; however, we also discovered that content, in and of itself, is rarely appealing to students. After careful consideration, we decided that our lesson plans needed to be recreated to match the present-day attitudes, interests, and perceptions of our students. The course content had to be personally relevant, socially shared, contentious, and interesting for students to fully appreciate the nature of science. As new teachers, our scientific knowledge from professional experiences provided insight regarding authentic contexts; however, we lacked the expertise in constructing a bridge to connect concepts and content to the world our students inhabit. Recognizing that becoming exemplary teachers would be problematic without learning how to provide scientific contexts in a format that would encourage students to *voluntarily* discover content knowledge and contemporary applications, we became graduate students in the College of Education at the University of South Florida (USF).

In most classroom discussions, the customary form of discourse is one in which the teacher asks the questions, a student gives a response, and then the teacher provides an assessment with commentary (Lemke 1990). This form of dialogue is not only severely constrained, but, on reflection, somewhat paradoxical. In the science classroom, questions are asked, surprisingly, not by those who *do not* know, but by the teacher who *does*. Many students are naturally reluctant to engage in such discourse in classrooms for fear of exposing their ignorance. In contrast, students need to realize that science is a relevant, cultural product where ideas need to be discussed and debated with peers to acquire perspective. In this environment, students discover the personal relevance of science as they share their individual beliefs and interpretations of specific content.

Graduate classes convincingly demonstrated that the contemporary science curriculum needs to include inquiries that encourage students to challenge and interrogate aspects of the scientific endeavor. There is a distinct difference between providing lectures and visual explanations of scientific phenomena and challenging students to discover content from contentious claims. Current media and political discussions focus upon contemporary issues that require the general population to have scientific understanding, such as performance-enhancing drugs, nutrition, genetic testing, and the validity of claims and evidence versus conspiracy theories. Students typically view science as difficult, boring, and irrelevant to everyday life (Lunetta 1998); therefore, they tend not to apply what they learn to life outside the classroom, and their naive misconceptions about phenomena in the natural world tend to remain unchanged (Roth 1990). The immediate challenge for teachers is to create a curriculum that includes contemporary issues and establishes guidelines so students understand the criteria of discerning reliable evidence, as well as develop methods of translating data into conceptual understanding. Because science is fundamentally about inquiry and analysis, posing questions about the world in which we live and then investigating and evaluating possible answers to those questions, encourages students to discover the content knowledge needed to challenge or substantiate explanations and beliefs.

Historically, content knowledge and correct contextual applications have been the focus of pedagogy studies. We dedicated an entire school year to designing curriculum and lesson plans, creating a model for implementing activities that focused on the use of socioscientific issues. Central to our purpose was to improve informal and moral reasoning through discourse, group activities, investigation of evidence, and argumentation. Contemporary SSI activities that were personally relevant, age appropriate and socially shared were selected to stimulate active inquiry of content knowledge and provide a framework for development of reasoning skills. There were many thoughtful discussions regarding the possible roadblocks to student acceptance of this pedagogy; however, we did not anticipate that the challenge of instituting a new curriculum design would also include the challenges of teacher transformation and confidence.

The following scenarios provide descriptions of using SSI in varied classroom environments, with diverse populations and subject matter. As teachers, we are expected to maintain extensive knowledge of a standardized curriculum and the textbook version of the subject; however, providing SSI curriculum requires new classroom management skills and contemporary knowledge for each SSI, as well as an understanding of the historic and contemporary frameworks of the contentious topics selected. While our coursework provided theoretical underpinnings for the SSI curriculum, learning the pedagogical practices necessary for the delivery of this approach was initially a difficult challenge. Without historical guidelines, continuous dialogue with professors provided a necessary level of confidence in situating SSI lesson plans.

The SSI Curriculum Framework: A Tale of Two Schools

Setting I: Palm Harbor University High School

Palm Harbor University High School is in a suburban, middle class neighborhood on the west coast of Florida. Most students transported from lower income areas of the county were African American; however, few African American students were enrolled in this particular SSI Project. The participants were members of four intact classes of 11th- and 12th-grade students (ages 16–18) enrolled in anatomy and physiology classes. Two of the classes were classified as "honors," made up of students who had excelled academically, and two of the classes were classified as "regular," made up of students with diverse histories of academic performance. Each of the classes contained between 29 and 31 students. One of each of the honors and regular classes was assigned to the comparison group; the other classes were assigned to the treatment group. The students were randomly assigned to the classes by school administrators; however, their placement within honors and non-honors sections was based upon the quality of grades earned in previous science classes and membership in the International Baccalaureate (IB) magnet program. A comparable number of students, including an equal number of males and females, were represented in both the honors and non-honors sections. Additionally, both treatment and standard curriculum honors sections contained equal numbers of IB students. It is significant to note that students enrolled in the IB magnet program had completed college level biology and chemistry courses. It is beyond the scope of this chapter to include a description of the grade transcripts of each student and class, but the teacher-observer noted that the rank of the students within their graduating class were comparable and equally distributed for all four classes.

SSI Curriculum Design

Figure 10.1 provides an overview of the curriculum created for this project. It reveals the inter-relationships between classroom science content knowledge and the complex social framework in which these concepts are embedded. More specifically, the figure illustrates the ten main SSI units, the primary anatomical systems related to the issues, and the connected scientific content and concepts developed by the corresponding arguments, debates, and discussion. The issues were carefully chosen to align with student interests and personal relevance, as well as specific science content standards. The lessons were designed to challenge core beliefs and apply new content knowledge to appropriate scientific contexts that are personally relevant. Topics included organ transplant allocation, the safety of marijuana and fluoridated water, the morality of stem cell research, euthanasia, quality of life issues, fast food consumption, and other contemporary subjects. Each SSI activity required three to seven class periods to complete; however, connections between context and content were reiterated throughout the academic year. The treatment curriculum was formulated to offer unique opportunities to confront, defend, or reject contentious information.

Both the comparison and treatment groups received explicit nature of science (NOS) instruction. The significance of NOS as a goal for science education is well established (Harding and Hare 2000; Irez 2006; Khishfe and Lederman 2006; McComas, Clough, and Almazroa 1998), so we took explicit NOS instruction as a given for science curriculum as well as a necessary

Figure 10.1. Socioscientific Issues Curriculum for Anatomy and Physiology

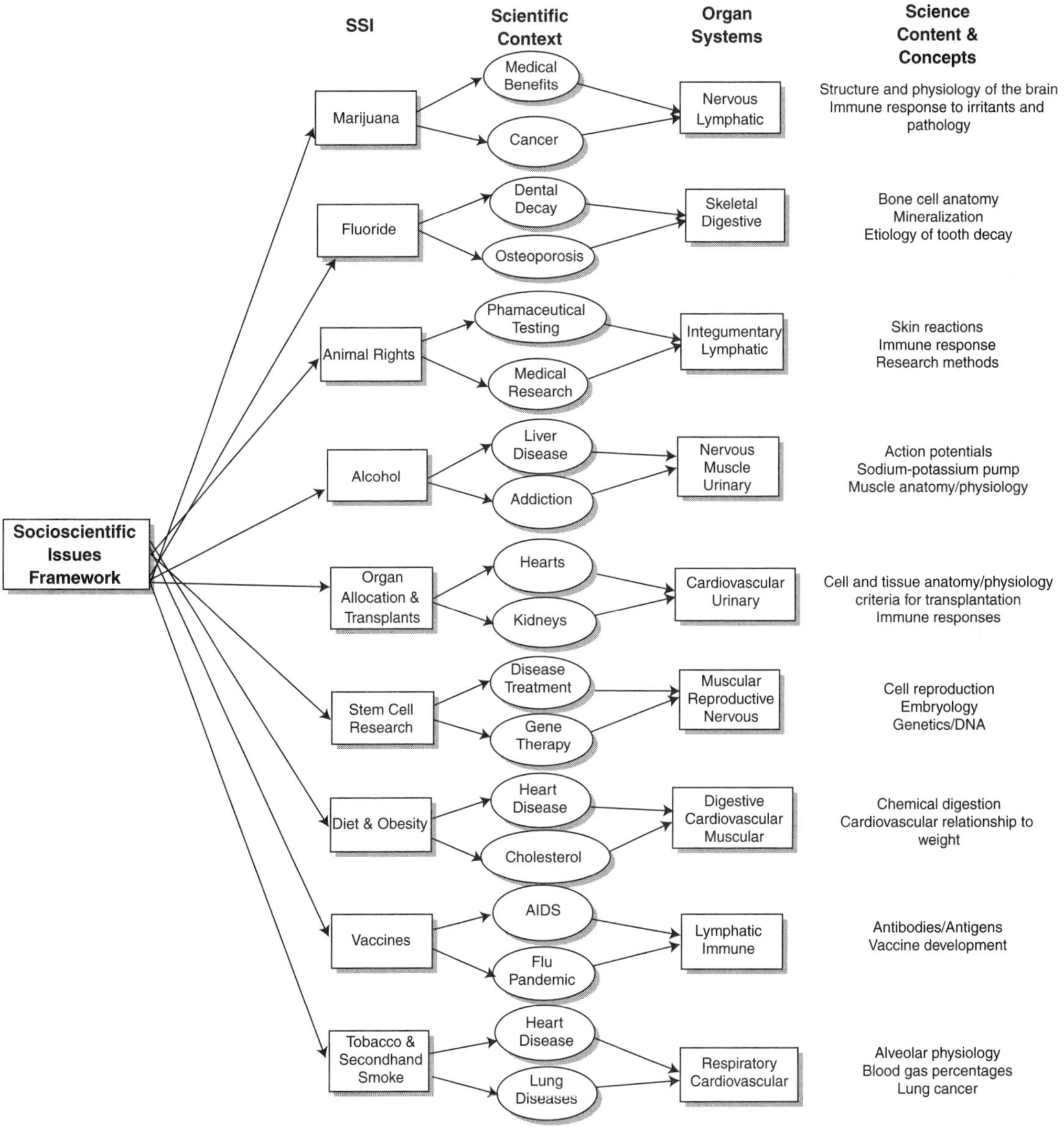

SSI	Scientific Context	Organ Systems	Science Content & Concepts
Marijuana	Medical Benefits / Cancer	Nervous Lymphatic	Structure and physiology of the brain Immune response to irritants and pathology
Fluoride	Dental Decay / Osteoporosis	Skeletal Digestive	Bone cell anatomy Mineralization Etiology of tooth decay
Animal Rights	Phamaceutical Testing / Medical Research	Integumentary Lymphatic	Skin reactions Immune response Research methods
Alcohol	Liver Disease / Addiction	Nervous Muscle Urinary	Action potentials Sodium-potassium pump Muscle anatomy/physiology
Organ Allocation & Transplants	Hearts / Kidneys	Cardiovascular Urinary	Cell and tissue anatomy/physiology criteria for transplantation Immune responses
Stem Cell Research	Disease Treatment / Gene Therapy	Muscular Reproductive Nervous	Cell reproduction Embryology Genetics/DNA
Diet & Obesity	Heart Disease / Cholesterol	Digestive Cardiovascular Muscular	Chemical digestion Cardiovascular relationship to weight
Vaccines	AIDS / Flu Pandemic	Lymphatic Immune	Antibodies/Antigens Vaccine development
Tobacco & Secondhand Smoke	Heart Disease / Lung Diseases	Respiratory Cardiovascular	Alveolar physiology Blood gas percentages Lung cancer

Socioscientific Issues Framework

component of a meaningful comparison treatment. NOS instruction was explicitly integrated throughout the learning environments of both the comparison and treatment groups. Our approach to NOS instruction was consistent with the field's dominant NOS framework (Lederman 2007) and involved several standard NOS activities (Lederman and Abd-El-Khalick 1998), including guided reflections on NOS themes and "black box" exercises in which scientific processes were modeled.

SSI Units of Inquiry

In Appendix A, page 164, we have selected one SSI unit (Marijuana Safety) that provides a closer view of the pedagogical details for using this topic as a controversial issue. Students were provided an opportunity to explore the drug's active ingredients and its effects on specific areas of the brain. Classroom debates and small and large group discussions encouraged students to confront their core beliefs about recreational drug use and personal thresholds of risk. In this particular activity, students were actively involved in reading and evaluating conflicting evidence from credible sources and negotiating their conclusions within and among other groups of students. This type of activity provided opportunities to observe the criteria students used in their selection of credible evidence. Students were required to work individually and in small groups, and to interact as a whole class. Their challenges included reading articles with conflicting evidence from varied sources, identifying important data and arguments, ranking the importance of evidence. They formed group consensus positions, debated positions, and served as editors for a mock scientific research journal with the goal of evaluating other groups' presentations of positions and evidence. Thus, the curriculum included multiple activities that required participants to evaluate claims, analyze evidence and their sources, come to a decision on a personal position, make moral decisions, and present the information within a group of peers to negotiate a consensus opinion.

The goal of student engagement in learning was essential to the success of the curriculum. Altering misconceptions was a major factor. Our working assumption was that SSI units afforded the context for students to understand, through carefully crafted experiences, that scientific knowledge is theory-laden and socially and culturally constructed. Cigarettes, alcohol consumption, recreational drugs, and steroid abuse were easily accepted topics; however, students were surprisingly attracted to debating issues that seemed peripherally relevant, such as fluoridated water, organ transplants, animal rights, and bacterial and viral epidemiology. An advantage of this format was the ability to adjust the themes to accommodate both the academic abilities and interests of the students. A concerted effort was made to select issues that would drive the anatomy and physiology curriculum, as well as to provide a context. Accordingly, the teacher's responsibility was to serve as a facilitator and guide.

Various measurements were collected to gauge understanding of fundamental scientific concepts using SSI as a framework (see Fowler, Zeidler and Sadler 2009; Zeidler et al. 2009); however, the scope of this chapter is intentionally limited to observations of qualitative changes in perception of SSI when challenging core beliefs and comprehension of science in contemporary contexts. Using 10 major SSI projects involving a minimum of 20 areas of scientific context provided students with multiple experiences (see Figure 10.1). Further, students were challenged to align their core beliefs with evidence supporting and/or opposing socioscientific issues. The teacher maintained a diary of student responses, including observations of gradual changes in individual and class understanding of science as personally significant. It seemed apparent from the first activities that the students reflected a deeper understanding of both scientific concepts and practical applications of content.

Summary of Diary Notations

Written notes were maintained after each activity, including descriptions of claims made and evidence used to substantiate assertions and develop understanding of the nature of science. Further, post-activity interviews were conducted to assess understanding of scientific concepts related to the designed activities.

Confronting Contextual Factors

When SSI were used as context, the content became relevant to students.

- Student ability to evaluate claims provided by media and other sources was improved when scientific concepts were related to relevant SSI.
  Example: Using the students' own personal observations and experiences regarding the use and abuse of alcohol, the failure of the muscular and nervous systems were introduced. Difficult concepts, including the movement of sodium and potassium across the phospholipid bilayer were learned with understanding, because it made sense in the perspective of muscle and nerve failure.
- Students demonstrated improved understanding of scientific concepts when they were able to attach the concept(s) to relevant SSI.
  Example: Local and national media attention of the case of Terri Schiavo provided the SSI background and an instructional opportunity to discuss the anatomical structure of the brain and the related physiology. Further, students were able to construct meaningful discussions on the various cultural definitions of life, dangerous diets, and the important of science in moral decision making.
- When presented with contemporary SSI, students were able to transfer conceptual understanding from one context and apply it to a new and different context.
  Example: Examination of the issue regarding stem cells, diseases of the nervous and muscular system, the effect of smoking on respiratory tissue, osteoporosis, and contagious diseases such as AIDS and influenza allowed students multiple opportunities to investigate cell structure and the driving principles of homeostasis, and the specific action of immune system mechanisms. Students demonstrated a better understanding of complementarity, the relationship between form and function, and science as the foundation for moral decision making.
- Students were able to manipulate variables (component parts) within a specific context to understand the direct and indirect effects on related concepts.
  Example: Students participated in role-playing activities during investigation of SSI, such as organ allocation, animal rights, and the matter of marijuana safety. The random

assignment of roles allowed students to challenge and defend their beliefs, using evidence they considered reliable. The use of various forms of evidence improved their skills in evaluating conflicting information.

Confronting Core Beliefs

Students were capable of evaluating and synthesizing data. However, when SSI provided information that was conflicting with their core beliefs, several interesting patterns emerged:

- Students often dismissed data (e.g., graphs, charts, and statistics) that were in conflict with their core beliefs, or failed to meet the criteria of personal experience.
  Example: The general population of students was motivated to participate in the research project, perhaps for the pure novelty of the circumstance. Early topics included water fluoridation and stem cell research because they were contemporary issues that had recently received local media attention. Students were provided summaries of articles that offered opposing viewpoints of the issues and they were instructed to write position compositions, based upon the evidence presented. All classes participated, but the treatment classes had received further instruction and discussion regarding argumentation and fallacious reasoning. The results were very surprising: The majority of students believed fluoride was harmful, ignoring substantial evidence that demonstrated 350 million people drank fluoride daily, without side effects or illness. The opposing article provided statements that indicated a possible link to cancers and dental disfigurement. The value of potential harm or negative consequences, even unsubstantiated, was more important than proven benefits.
- The perceived value and relevance of information was based upon its fit with personal experiences.
  Example: Current media assertions by nonscientist authorities (government leaders) proclaimed that stem cell research was comparable to abortion. Without personal experience in areas of illogical reasoning, the students reverted to fundamental, core beliefs and expressed a genuine fear of possible illegality and religious sin.
- Academic understanding failed to provide confidence for students to trust their reasoning skills. It was early in the curriculum and the teacher was losing confidence in the goals of the project. The socioscientific issue was contemporary, but not personally relevant.
  Example: Because the students are in their "immortal" years, enormous amounts of evidence demonstrating the connection between unhealthy diets, smoking, and heart disease seem only remotely relevant. However, they are becoming more astute about evidence and have focused their attention on data that proves the age of inception of cardiac disease has dropped from age 52 to 38, based upon the increase in fast food consumption. In this context, students learned the anatomical and physiologic changes related to cardiac pathology.
- When students were compelled to defend their opinions to their respective group, the class, and the teacher, they included their core beliefs and personal experiences in their defenses.

Example: In defending animal rights, students were required to obtain multiple perspectives, including the activities of medical researchers and federal regulators. When students were forced to defend a position that was not consistent with their personal beliefs, it provided an opportunity to challenge the credibility of their opinions that had been developed entirely around the love of the family pet and animal shows on television. Arguments had to be substantiated with evidence, demonstrating that science requires empirical data.

- Students were generally surprised that reliable sources of scientific information at times provide conflicting claims and conclusions.
  Example: Student involvement in weekly discussions and regular SSI projects has resulted in comfortable group negotiation of scientific issues and evidence to develop a consensus opinion. In an activity based upon conflicting evidence, in regard to the potential harms and benefits of marijuana, students evaluated various sources of data and information authorities. At the end of the yearlong curriculum approach, it became apparent that the students were able to express opinions about concepts and evidence while simultaneously challenging their own belief system.

While many students' personal belief systems remained intact, many were able to include compatible science concepts in a new belief system. The greatest achievement was observing development of a more mature attitude in the formation of a consensus resolution to dilemmas even when their personal beliefs conflicted with the decision of their respective group. Science teacher educators must understand that the process of challenging and perhaps rejecting deeply held beliefs is extremely difficult. Many activities elicited a great deal of anxiety, which resulted in classroom discourse where personal values were heatedly discussed and questioned. Teacher transformation included the discovery that when implementing an SSI curriculum, it is imperative to maintain a learning environment conducive to the safe expression and exploration of ideas and thoughts by individuals and groups.

Setting II: Robinson High School

The students were enrolled in four classes of high school marine science (biology) at Robinson High School, an urban public high school located in Tampa, Florida. Most students who attend Robinson High School are from middle to lower socioeconomic families. The school operates on a Copernican four-by-four block schedule, with 90-minute classes. The participating classes consisted of 125 students, grades 10 through 12, ages 15 through 18. The students represent a wide range of socioeconomic and ethnic origins; specifically, African American (10), Hispanic (26), Caucasian (81) and Asian (8). Six students spoke English as their second language. It was determined that 52% the students were primarily kinesthetic learners, 28% were visual learners, and 20% were auditory learners. Approximately 16.7% were children from single female household (with no husband present), and 9.55% with children under the age of 18. Forty-six percent of students qualified for the free and reduced lunch program. Robinson High School has a well-diversified population of 1200 students, including 182 Exceptional Education (ESE) students and children from military families assigned to MacDill Air Force Base.

Three SSI units were included in lesson plans. The first case study, *Oil Spill in Tampa Bay, What a Mess!* was presented as a part of a unit on human impact on the environment. In this scenario, a hurricane entering Tampa Bay caused an oil tanker to run aground on a sand bar. The hull of the oil tanker ripped open, spilling thousands of gallons of crude oil into the Tampa Estuary. During the SSI unit, students were expected to investigate the impact of oil including aspects of the spill, contamination, and our dependence on oil. Students engaged in an inquiry lab to determine the best method of cleanup after an oil spill. Next, they participated in role-playing as members of a hypothetical town where citizens are obligated to decide how to clean up the oil, how to fund the clean-up, and whether to allow oil tankers to enter Tampa Bay in the future. Further, if oil delivery is permitted, the students constructed laws and guidelines for better oil tanker engineering, to prevent future oil spills.

The second SSI unit, *Red Tide: A Community in Crisis,* was organized as a part of a larger unit on the study of marine algae. In this scenario, Seaside City, a fictitious town in Florida has suffered an outbreak of red tide during the summer tourist season. Numerous individuals have been suffering from respiratory symptoms believed to be associated with red tide, and the destructive algal bloom is keeping tourists away from the town and its beaches. During this SSI unit, students were expected to learn the characteristics of different algae, including diatoms, dinoflagellates, and multicellular algae, as well as marine flowering plants. They participate in inquiry labs to determine structure, characteristics, and geographic locations of various forms of algae. Students also investigate harmful algal blooms (HABs), specifically Florida's red tide known as *Karenia brevis*. Finally, students assume the roles of various characters having vested interests in the scenario in order to critically reason and debate issues about red tide.

The final SSI: *Beach Nourishment: A Community Crisis*, was included as part of a unit on marine conservation. In this scenario, Seaside City, a fictitious town in Florida, has suffered major beach erosion over the past several years. Citizens have been asked to vote to close down the beaches or financially support restoration of the beaches due to the importance to tourism. Students are expected to investigate the processes of beach erosion, describe and evaluate the various types of beach restoration, and examine the challenges and impact of planning the implementation of a beach restoration plan. The culminating activity was a role-play where students were assigned characters to examine beach restoration from multiple perspectives and critically reason and debate issues about the management of beach erosion and restoration.

The SSI units were introduced in the treatment class each semester, while the other class was taught the content using traditional pedagogies of lecture, prescribed laboratory experiments, and multiple-choice assessments.

Summary of Diary Notations

Confronting Contextual Factors

When SSI were used as context, then the content became relevant to students.

- Student ability to evaluate claims provided by media and other sources was improved when scientific concepts were related to relevant SSI.

Example: Media clips from local news stations and reports from a local scientific agency introduced the health hazards of red tide. Students referred to their personal observations and experiences when they went to the beach during outbreaks of red tide along Florida's Gulf Coast. They discussed how they coughed, suffered from red itching eyes, and felt their throats burning. They also referenced the dead fish they observed washed up on shore. Making connections between brevitoxins produced by *Karenia brevis* and their impact on human health became meaningful due to personal experiences.

- Students demonstrated improved understanding of scientific concepts when they were able to attach the concept to relevant SSI.

  Example: Examination of the restoration efforts of a local beach in Pinellas County, Florida allowed multiple student opportunities to investigate beach dynamics, beach geology, interstitial communities, local food webs, and ecological impact of human activities. Students undertook meaningful discussions regarding the various definitions and characteristics of a healthy beach, and the importance of science in moral decision-making with regards to publicly owned lands.

- Students were able to transfer conceptual understanding from one context and apply it to a new or different context.

  Example: National media attention to oil leaks and spills from various shipwrecks provided the SSI background and instructional opportunity to discuss the impact of oil spills on animal health and the ecosystem. Additionally, through inquiry investigations into oil spill cleanup, students were able to construct meaningful arguments for or against different methods. Examination of the risk and benefit analysis of oil tankers in Tampa Bay allowed students to understand the relationship between science and political decision making with regards to laws and regulation.

- Students were able to manipulate variables (component parts) within a specific context to understand direct and indirect effects on related concepts.

  Example: Student role-playing during the negotiation of an SSI allowed them to examine environmental issues from multiple perspectives. Students drew their roles from a hat, allowing for random selection of viewpoints. Student opportunities to challenge and defend their beliefs, using evidence they deemed reliable. Using various forms of evidence (e.g., scientific data, personal experience, emotive core beliefs), students improved their skills in evaluating conflicting information and argumentation.

Confronting Core Beliefs

Students were capable of evaluating and synthesizing data. However, when SSI provided information that was conflicting with their core beliefs, several interesting patterns emerged:

- Students often dismissed data that were in conflict with their core beliefs, or failed to meet the criteria of personal experience.

  Example: In general, students from all classes were motivated to research course topics, whether presented as an SSI scenario or as a content-specific concept. This general interest across groups may have been due to the emotive nature of topics. Students were asked to create brochures that summarized concepts related to contemporary marine environmental issues, and to write opinions or take positions based upon the evidence presented. All classes participated, but the treatment classes received further instruction in argumentation and decision making. In the case of the oil spill, despite the different instruction provided to the classes, the majority of students believed that oil tanker spills were the main source of oil in the marine environment, despite being presented with graphs and data demonstrating that natural oil leaks accounted for more oil in the marine environment than any other source. The risk of an oil spill and the potential harm to the ecosystem was more important when deciding to not allow oil tankers in Tampa Bay to deliver crude oil.

- When students were compelled to defend their opinions to their respective groups, to the class, and to the teacher, they included their core beliefs and personal experiences in their defenses. *Example:* In determining whether or not to implement a beach restoration plan, students were required to include multiple perspectives, including the interests of hotel owners, tourists, local residents, and environmentalists. Defending a position that did not align with their personal beliefs provided an opportunity to challenge the credibility of their opinions, which were often developed from their personal experience as a beach user. Arguments had to be substantiated with scientific evidence and empirical data. In these cases, students often referred to the authority of evidence using statements such as "Dr. X from the U.S. Geologic Survey…." Yet, students would follow up this argument with a narrative of a personal experience whether or not it was relevant. For example, "But when I was at the beach last weekend, it was too dangerous to surf due to the rock jetties."

Conclusions

The previous assertions are not intended to discount the value of more than 30 years of curriculum effort and publications by NSF, AAAS, and other organizations regarding STS; however, recent academic and educational research has demonstrated that effectively teaching scientific concepts needs contemporary relevance. Creating the mantra of rigor and personal relevance in science classrooms is insufficient unless teachers can use the investigation of socioscientific issues as a context for discovering underlying scientific concepts. It has been determined that the process of developing scientific knowledge is enhanced by discovery and active exploration of socioscientific issues. Many educators have encouraged students to examine conflicting evidence, negotiate personal perspective, and challenge their core beliefs about contentious scientific topics; but, this approach is not often currently a part of a standardized curriculum. Sociomoral discourse, argumentation, and debate have been clearly established as necessary elements in character development and decision-making abilities and, therefore, should be essential components of science education. While this pedagogy requires students to become actively engaged in socially shared activities that "unearth" personal connections and relationships to contentious scientific topics, it is imperative that teachers possess the characteristic leadership and teaching skills requisite to guide students in their exploration and understanding of science. It is not unreasonable to provide specific instruction on the subject of transforming teacher/subject matter-centered classrooms to student/issues-based investigation communities in education methods classes.

Science that is encountered by most people in their everyday lives is rarely objective, coherent, well bounded, and unproblematic. Science beyond the laboratory turns out to be a messy business and, far from scientific information being central to decisions about practical actions, it is often irrelevant, or at best, marginal to it. Scientific knowledge is not encountered free of its social and institutional connections and scientific thought is often rejected as the yardstick by which to judge the validity of everyday thinking. In addition, ignorance of science on the part of adults may have nothing to do with intellectual ability; such ignorance may not always be seen as in need of remedy. The significance of this account of science is that it stands firmly at odds with the view of science commonly experienced at school, where the discipline is well bounded; answers are secure; and uncertainty, doubt, and debate are not admitted. School science often reflects an endeavor that is essentially positivist, heroic, apolitical, and more concerned with scientists of the past than with those of the present. It can be argued with considerable justification that much (almost all) of school science is unproblematic. Accepting this, however, leaves students confronted with two seemingly conflicting and overlapping visions of science: One is constructed and institutionalized in the school curriculum, and the other is much less certain and develops from their own, rapidly enlarging, experience of the social, physical, and emotional worlds which they inhabit. We should not be surprised, therefore, if the understanding of scientific phenomena promoted at school rarely transfers to the solution of problems in the out-of-school world.

Use of an SSI curriculum requires the classroom instructor to possess the knowledge of scientific concepts, contemporary applications and their relationship to present-day discussion and events familiar to students. Teachers then must urge students to "discover" a relationship between personal relevance and scientific data. Whereas teaching proficiency normally requires experiential knowledge from textbooks, science instructors need to implement this unique pedagogy by acquiring specific and peripheral information from internet sources, scientific journals, popular magazines, newspapers, and television. In this regard, the first years of employing SSI as a context for learning scientific content requires a commitment to discovery and data collection by teachers. Teacher content knowledge and the ability to determine the basis of certain scientific issues needs to be considered before implementing SSI-based instruction. Because argumentation is a fundamental process of negotiating issues, it is helpful for teachers to be familiar with methods of assessing the quality of arguments (Gess-Newsome and Lederman 1995; Gess-Newsome 1999). Finally, the epistemological orientation and core beliefs of teachers need to be examined, with regard to their understanding of the development of scientific knowledge. It would not be unrealistic to find that many teachers still believe that scientists only follow the scientific method to generate knowledge; often their understandings are limited to textbook explanations. Decision making that requires evaluation of claims and evidence, as well as discerning connections between concepts, and context requires more than science content knowledge.

Is global warming a product of industrial negligence? Is marijuana harmful? Is stem cell research an ethical practice? Teachers must possess the pedagogical skills to create and direct activities that introduce the subject matter as contemporary, personally relevant, contentious, and reaffirming or challenging core beliefs. For example, an activity that encourages learning DNA and the mechanism of genetics would include investigation, discourse, and argumentation about the genetic predisposition of gender preference. When contentious issues are examined, students are provided the opportunity to exercise reflective judgments, develop scientific literacy, and practice decision-making skills. Within the SSI framework, students are exposed to moral problems that involve a number of discrepant scientific, social, or moral viewpoints, many of which may conflict with the student's own closely held beliefs. Using issues-based curriculum, teachers are compelled to provide activities that demonstrate that scientific knowledge is not absolute, but forms as a result of social knowledge construction from argumentation and discourse. These new goals require teachers to transform their pedagogical orientation from being purveyors of scientific knowledge to moderators and mediators of a classroom culture that mirrors society in which students are challenged to make informed scientific decisions and exercise moral reasoning.

Effective use of an issues-based curriculum requires teachers to understand subject matter deeply and be flexible so they can assist students in the creation and scaffolding of driving principles and core conceptions. The objective is for teachers to transform their pedagogical focus and scientific epistemology so students can better understand how such knowledge is generated and validated (Abd-El-Khalick 2006). Within the SSI framework, students are exposed to moral problems that involve a number of discrepant scientific, social, or moral viewpoints, many of which may conflict with the student's own closely held beliefs. Using an issues-based curriculum compels teachers to provide activities that demonstrate that scientific knowledge is not absolute, but forms as a result of social knowledge construction from argumentation and discourse.

The success of using an SSI-based curriculum is contingent upon redefining the role of the teacher and the responsibilities of the students. A classical classroom setting includes students sitting apart, in rows, and the instructor standing behind a lectern or in front of a PowerPoint presentation of notes and diagrams. Believing that learning is a socially shared activity, SSI projects are deliberately designed to remove the teacher from the focal point of the classroom; the activities require students to independently develop claims and negotiate conclusions within their own groups. Zeidler and Keefer (2006) have proposed that the role of the teacher should include development of inquiry, discourse, and argumentation skills because science issues require effective decision making. Understanding that the essence of SSI instruction means acquiring a tacit awareness of knowledge about beliefs concerning the role of context in instruction (e.g., knowledge of students, community, school, culture) suggests a transformative model is needed for effective teaching.

Because SSI instruction introduces issues with moral dilemmas, conflicting evidence, as well as multiple sources of evidence, teachers are expected to evaluate claims regarding the sources of information provided by students. Student presentation of written and oral arguments and conclusions transforms the role of the teacher from being the sole source of information into that of a mediator and moderator. As part of the transformation process, teachers need to develop competence in areas of critical thinking, argumentation, quality assessments, and frequent discussion of moral dilemmas. Adaptations to this new curriculum requires the teacher to transform his/her attitude about being the singular source of information to encouraging their students to make individual decisions, even when personal beliefs are mistaken for scientific concepts. Recent studies have demonstrated that when SSI is used as science pedagogy, students can handle conflicting evidence by drawing upon past experiences and combining them with new information, to explain actions in a scientific context (Driver et al. 1996; Driver, Newton, and Osborne 2000; Kolstø 2001; Sadler 2004; Zeidler and Keefer 2006; Zeidler and Sadler 2008).

Implications: Transformative Ideas for Science Teacher Education

Teachers need to transform their pedagogical orientation away from introducing science concepts through simple lectures and reconfirming laboratory investigations; instead, teachers need to create classroom environments where students can develop meaningful understandings of scientific concepts in relationship with real-world circumstances. Student learning transformation occurs when they extend their interpretations of concepts and make connections with other situations in their daily lives.

Science teacher education is primarily concerned with providing viable frameworks that teachers can utilize to engage students in the activity of science and develop meaningful (functional) notions of scientific literacy. For preservice and practicing teachers, the realization that science education for many (most) students has included years of indoctrination, dogmatism, and authoritarianism is a sobering epiphany. However, there is no place in science and, therefore, no place in science education, for the protection of concepts and theories from criticism. The challenge for science teachers is to allow students to have personal experiences that do not immediately negate their belief systems; rather, the aim is to provide the conditions necessary to enable the development of a personal epistemology through continued exposure to, and interaction with, the nature of science and SSI. The use of argumentation and relevant SSI as a framework for science class curricula is essential for enabling scientific concepts to enter individual belief systems of students. Experiences in life and experiences in teaching over time may help one's comfort level. However, the importance of taking pedagogical risks demands that teachers trust in student ability to extract content knowledge from SSI activities. The teacher's role transforms to one of facilitation rather than final authority.

Students are capable of learning when the information is part of contemporary and relevant socioscientific issues that must be actively investigated to be successfully argued and negotiated within groups. Global warming, ozone depletion, and other environmental issues are important and relevant, but the "arms length" distance to their daily lives and the beliefs of students that their opinions have little, if no direct effect or value, creates a barrier for teaching and learning. If possible, the teacher should investigate an issue that is closer to the students' daily existence. Sadly, drugs and alcohol are easy and relevant. To bring powerful ideas into focus, you need to find what significantly offends, cajoles, provokes, and makes what otherwise may be distant topics central to the everyday lives of students. Only then can such ideas make a classroom authentically transformative.

The fatal flaw held by many teachers is their own pedagogical belief that concepts can be taught using sufficient explanations and tidy analogies that will then magically alter the core beliefs of students. The use of classroom activities, such as black box activities and argumentation, challenges students to not only re-evaluate their prior understandings, but also provides an opportunity for them to restructure their conceptual understanding of subject matter through personal experiences and social discourse. Smokers quit their addiction when they get cancer and people exercise and diet with a sense of urgency having experienced a heart attack. In both cases, the experience has been transformative in the sense that the source of their newfound motivation comes from an experience that is at once personal and relevant. For students to change their epistemological beliefs about scientific data, the educative experience must too be personal and relevant.

References

Abd-El-Khalick, F. 2006. Socioscientific issues in pre-college science classrooms. In *The role of moral reasoning and discourse on socioscientific issues in science education*, ed. D. L. Zeidler, 41–61. Dordrecht, Netherlands: Springer.

Driver, R., J. Leach, R. Millar, and P. Scott. 1996. *Young people's images of science*. Bristol, PA: Open University Press.

Driver, R., P. Newton, and J. Osborne. 2000. Establishing the norms of scientific argumentation in classrooms. *Science Education* 84 (3): 287–312.

Fowler, S. R., D. L. Zeidler, and T. D. Sadler. 2008. Moral sensitivity in the context of socioscientific issues in high school science students. *International Journal of Science Education* 31 (2): 279–296.

Gess-Newsome, J. 1999. Pedagogical content knowledge: An introduction and orientation. In *Examining pedagogical content knowledge*, eds. J. Gess-Newsome, and N. G. Lederman, 147–163. Boston: Kluwer Academic Publishers.

Gess-Newsome, J., and N. G. Lederman. 1995. Biology teachers' perceptions of subject matter structure and its relationship to classroom practice. *Journal of Research in Science Teaching* 32 (3): 301–325.

Harding, P., and W. Hare. 2000. Portraying science accurately in classrooms: Emphasizing openmindedness rather than relativism. *Journal of Research in Science Teaching* 37: 225–236.

Irez, S. 2006. Are we prepared? An assessment of preservice science teacher educators' beliefs about nature of science. *Science Education* 90: 1113–1143.

Khishfe, R., and N. G. Lederman. 2006. Teaching nature of science within a controversial topic: Integrated versus nonintegrated. *Journal of Research in Science Teaching* 43 395–318.

Kolstø, S. D. 2001. "To trust or not to trust. . ." pupils' ways of judging information encountered in a asocioscientific issue. *International Journal of Science Education* 23: 877–901.

Lederman, N. G. 2007. Nature of science: Past, present, and future. In *Handbook of research on science education*, eds. S. K. Abell and N. G. Lederman, 831–880. Mahwah, NJ: Lawrence Erlbaum Associates.

Lederman, N. G., and F. Abd-El-Khalick. 1998. Avoiding de-natured science: Activities that promote understandings of the nature of science. In *The nature of science and science education: Rationales and strategies*, ed. W. F. McComas, 83–126. Dordrecht, Netherlands: Kluwer.

Lemke, J. L. 1990. *Talking science: Language, learning and values*. Norwood: Ablex Publishing.

Lunetta, V. N. 1998. The school science laboratory: Historical perspectives and context for contemporary teaching. In *International Handbook for science education*, B. Fraser and K. Tobin, eds., 249–262. Dordrecht. The Netherlands: Kluwer.

McComas, W. F., M. P. Clough, and H. Almazroa, (1998). The role and character of the nature of science in science education. In *The nature of science in scienceeducation*, ed. W. F. McComas, 3–39. Dordrecht: Kluwer Academic.

Roth, K. J. 1990. Developing meaningful conceptual understanding in science. In *Dimensions of thinking and cognitive instruction*, eds. B. F. Jones and L. Idol, 139–175. Hillsdale, NJ: Erlbaum.

Sadler, T. D. 2004. Informal reasoning regarding socioscientific issues: A critical review of research. *Journal of Research in Science Teaching* 41: 513–536.

Walker, K. A., and D. L. Zeidler. 2007. Promoting discourse about socioscientific issues through scaffolded inquiry. *International Journal of Science Education* 29 (11): 1387–1410.

Zeidler, D. L. 2002. Dancing with Maggots and Saints: Past and future visions for subject matter knowledge, pedagogical knowledge, and pedagogical content knowledge in reform and science teacher education. *Journal of Science Teacher Education* 13 (1): 27–42.

Zeidler, D. L. 2003. *The role of moral reasoning and discourse on socioscientific issues in science education.* Dordrecht, Netherlands: Kluwer Academic Publishers.

Zeidler, D. L., S. Applebaum, and T. D. Sadler. 2006. Using socioscientific issues as context for teaching content and concepts. Paper presented at the Annual Meeting of the Association for Science Teacher Education, Portland, OR.

Zeidler, D. L., and M. Keefer. 2006. The role of moral reasoning and the status of socioscientific issues in science education: Philosophical, psychological, and pedagogical considerations. In *The role of moral reasoning and discourse on socioscientific issues in science education*, ed. D.L. Zeidler, 7–38. Dordrecht, Netherlands: Springer.

Zeidler, D. L., and T. D. Sadler. 2008. The role of moral reasoning in argumentation: Conscience, character and care. In *Argumentation in science education: Perspectives from classroom-based research*, eds. S. Erduran and M. Pilar Jimenez-Aleixandre, 201–216. New York: Springer Press.

Zeidler, D. L., T. D. Sadler, S. M. Applebaum, and B. E. Callahan, 2009. Advancing reflective judgment through socioscientific issues. *Journal of Research in Science Teaching* 46 (1): 74–101.

Appendix A—Teacher Overview of Marijuana SSI Activity

Marijuana Activity Outline

<u>Overview:</u> The focus of this activity is student negotiation of the controversy surrounding the physiological effects of marijuana. The activity challenges students to consider data representative of multiple perspectives relative to the issue of physiological effects of marijuana. Students review five articles from a variety of sources presenting disparate lines of evidence supporting and refuting significant medical impacts resulting from the use of marijuana. The students then work in small groups to prioritize pieces of evidence drawn from the articles reviewed. In the final phase, each small group evaluates how a peer group prioritized the data. The sections that follow provide detailed descriptions for each step of the activity.

<u>Phase I:</u>
<u>Goal:</u> Students become familiar with five articles representing various perspectives relative to the issue of the physiological effects of marijuana use. Students isolate pieces of evidence used to support positions presented in the articles.

<u>Materials:</u> Articles (contained in 1 packet); Phase I Activity Sheets (1/student)

<u>Grouping:</u> Students work on this phase *individually*.

<u>Procedure:</u> Each student reads all five articles. For each article, the student must 1) Summarize the main argument or position of the article; 2) Identify all lines of evidence (relative to the physiological effects of marijuana use) presented in the article; 3) Identify the source of the article.

   After completing these tasks for all five articles, students are asked to answer two questions based on the whole set: 1) Which article do you find most convincing? Please explain. 2) Which article has the most scientific merit? Please explain.

<u>Student Product(s):</u> Phase I Activity sheets, from each student, which provide written documentation of the evidence they have identified for each article.

<u>Phase II:</u>
<u>Goal:</u> Students evaluate and prioritize the evidence presented in the five articles.

<u>Materials:</u> Each student should use his/her completed Phase I Activity Sheet; Phase II Activity Sheet (1/student)

<u>Grouping:</u> Small groups of 3-4 students

Procedure: Small groups of students come together and must collectively determine the six most important pieces of evidence presented across the five articles. This process should involve group negotiation, prioritization of all the evidence, and justification of their final ranking.

When a group reaches consensus on the most important lines of evidence, each student must write a justification for their group's final ranking.

When all members of the group complete their written justifications, they should select a single justification which best represents their negotiations.

Student Product(s): Phase II Activity Sheets, from each student, which will provide the group ranking of evidence and a written justification (a paragraph or two) for why his/her group chose to prioritize its chosen lines of evidence over others.

Phase III:

Goal: Students will critically evaluate how another group prioritized evidence from the five articles.

Materials: Each group will receive the evidence rankings and "best justification" from another group; Phase III Activity Sheet (1/group); *Palm Harbor Journal of Scientific Research* review letter (1/group)

Grouping: Students should work in the same groups formed for Phase II.

Procedure: The premise of this phase is that the students are serving as reviewers for the *Palm Harbor Journal of Scientific Research (PHJSR). PHJSR* seeks to publish brief summaries of evidence relative to the physiological effects of marijuana use. The group must review the six pieces of evidence chosen by their peer group as the most important and the accompanying justification. The group must then accept or reject this work for inclusion in the PHJSR. As in all journal reviews, the reviewers must explain their decision in writing (one or two paragraphs).

Student Product(s): Each group will complete the PHJSR review letter, which will record its decision (accept/reject) and provide a written explanation for its decision.

Appendix B—Three Phase SSI Activity Sheets for Students

Exploring the Physiological Effects of Marijuana
Phase I Activity Sheet

Directions: You have been given five articles related to the physiological effects of marijuana. For each article, you will need to: (1) briefly summarize the main argument or position of the article; (2) identify all of the evidence used in the article in order to support the position; and (3) note the article's source (in other words, write down who is responsible for the article and the research presented. Keep in mind that a single article may (or may not) present multiple pieces of evidence.

Use the summary sheets for each article attached to this packet.

Article 1: Heavy marijuana use linked to brain damage

Briefly summarize the main argument or position of the article:

_____

_____

Identify evidence used in this article:

1)_____

2)_____

3)_____

4)_____

5)_____

Identify the source of the article: _____

Article 2: No brain damage seen in marijuana-exposed monkeys

Briefly summarize the main argument or position of the article:

_____

_____

Identify evidence used in this article:

1)_____

2)_____

3)_____

4)_____

5)_____

Identify the source of the article: _____

Article 3: Researchers report smoking marijuana may increase risk of head and neck cancers

Briefly summarize the main argument or position of the article:

_____

_____

Identify evidence used in this article:

1)_____

2)_____

3)_____

4)_____

5)_____

Identify the source of the article: _____

Article 4: Marijuana: A scientific controversy

Briefly summarize the main argument or position of the article:

_____

_____

Identify evidence used in this article:

1)_____

2)_____

3)_____

4)_____

5)_____

Identify the source of the article: _____

Article 5: Turn on, tune in, get well

Briefly summarize the main argument or position of the article:

_____

_____

Identify evidence used in this article:

1)_____

2)_____

3)_____

4)_____

5)_____

Identify the source of the article: _____

Exploring the Physiological Effects of Marijuana
Phase II Activity Sheet

Directions: <u>Phase II</u> should be completed in small groups of 3–4 students. The goal of this phase is to identify and rank the most important pieces of evidence presented across all five articles related to the physiological effects of marijuana. Keep in mind that you are being asked to rank the MOST IMPORTANT PIECES OF EVIDENCE; you are not asked to rank whole articles. Groups should initially work as a team to agree on the six most important pieces of evidence and then rank your final choices. (Evidence listed as #1 is the most important piece of evidence; evidence listed as #2 is the next most important piece; etc.) *Each* group member should record the final group rankings and then write a one- or two-paragraph description of why you and your group ordered the evidence as you did. Every student should complete a Phase II Activity Sheet.

Evidence Rankings (*Identify and rank the six most important pieces of evidence relative to the physiological effects of marijuana.*)

1. _____

2. _____

3. _____

4. _____

5. _____

6. _____

Justification. *In the space below write a few paragraphs describing how and why you and your group chose the 6 lines of evidence as the most important pieces of evidence related to the physiological effects of marijuana from all of the material that you read. You should probably focus on what made the items you chose different from items that were left off your list. You should also discuss why the listed items are ranked the way they are (e.g., why is evidence #1 more important than evidence #2, etc.) You may also discuss reasons why evidence presented in the articles but not included on your list was left out. Each group member should write his/her own justification, but you are reporting on the work you did as a group.*

Exploring the Physiological Effects of Marijuana
Phase III Activity Sheet

Directions: In this final phase of the activity, your group will serve as a panel of reviewers for the *Palm Harbor Journal of Scientific Research (PHJSR). PHJSR* wants to publish brief summaries of evidence relative to the physiological effects of marijuana use. Like all scientific journals, *PHJSR* requires that experts rigorously review the information printed.

Your task will be to review (as a group) the suggestions from another group on the most important pieces of evidence in the controversy related to the physiological effects of marijuana use. To do this, you should carefully consider the chosen evidence, how that evidence is ranked, and the quality of the justification.

Ultimately, you must accept or reject the other group's work for inclusion in the *PHJSR*. You will communicate your decision to the editors by completing the *PHJSR* review letter, which includes an area for you to accept or reject and an area for you to write an *explanation* (one to two paragraphs) for your decision. You should work on this task as a group and submit one final product. (Use the format on the next page for your journal review letter.)

Palm Harbor Journal of Scientific Research

Editorial Review Form

Dear Editors:

As members of your review team, we have completed our review of the data for submission to *PHJSR*. After careful consideration, we recommend that this work (will) (will not) be accepted for publication in our prestigious journal. The reasons for our decision are indicated below.

Quality of Evidence Selected

Quality of Evidence Ranking

Quality of Justification for Evidence Selected and Ranked

Securing a "Voice":

The Environmental Science Summer Research Experience for Young Women

David L. Brock
Roland Park Country School

Setting

Roland Park Country School is an urban college preparatory school for students in grades K–12 located in Baltimore, Maryland. An independent all-girls school, the institution maintains a commitment to community outreach to address issues of equity and social justice, and as part of this investment, we have made environmental education and stewardship a central part of our school mission. Five acres of heavily wooded area on our campus were set aside for permanent Forest Conservation Easement in 2001, and for nine years now, a team of young women from the greater metropolitan area have studied and analyzed the microclimates in this preserve to aid the school in its conservation and environmental outreach efforts. Known as the Environmental Science Summer Research Experience for Young Women (E.S.S.R.E.), this three-week summer internship runs every July.

Women in Science: Addressing a Need

As a school dedicated to single-sex education, our raison d'etre is to nurture and empower young women to pursue their intellectual passions. In particular, we are committed to encouraging girls devoted to science and helping them overcome the stereotypical barriers still confronting them today. Research by the American Association of University Women (AAUW 2004) has shown that women continue to face discrimination and discouragement at all levels of their education in the sciences, and in spite of recent gains in the numbers of PhDs awarded to women in the sciences, the proverbial glass ceiling that exists at most universities and other research institutes still remains a potent barrier to the participation of women in the sciences (Committee on Science, Engineering, and Public Policy 2007). Moreover, changes in this status quo have been extremely slow in coming (AAUW 2004); hence, the need for "providing challenging opportunities for all students to learn science" (NRC 1996) remains as imperative now as it did when the current science education reform efforts began over a decade ago.

A major goal in creating E.S.S.R.E. was to address this need. By providing just the sort of apprenticeship teenage girls need to cultivate their instinctive adolescent curiosity, the program strives to sustain their interest in the sciences at precisely a decisive stage when this interest can be lost from neglect. E.S.S.R.E. is the only internship of its kind for 9th- and 10th-grade girls in the state of Maryland, and in its interface of research experience and environmental stewardship, the program seeks to promote awareness and ownership of scientific investigation and its consequences that enables girls to have a more active "voice" in science. Put simply, critically low numbers of women participate in the sciences in any way at all in our society, and therefore, E.S.S.R.E. endeavors to produce scientifically literate girls who can participate more knowledgeably in all facets of their communities. We are always quite pleased when one of our interns goes on to pursue a specific career in the sciences, but what our program ultimately seeks is to generate women who understand that science and its unique way of looking at the world is theirs to employ in meeting life's critical needs.

The Program

Participants in E.S.S.R.E. study soil ecology and perform self-designed field study investigations in order to accomplish several major goals:

1. to develop a better understanding of key theoretical principles and the practical application of these to a changing ecosystem;

2. to aid in the management of our school's campus and the conservation of its wooded areas;

3. to raise awareness of critical issues and concerns facing the larger environment in which our school resides;

4. to promote and improve environmental science education in schools everywhere; and

5. to learn applicable skills that can transfer to other research situations and academic fields.

The girls meet these goals by performing a number of different tasks. To begin, they work in teams of three or four to complete a modified version of the Smithsonian Institution's biodiversity assessment method (National Zoological Park 2009). The groups measure 20 × 20 m quadrants in their assigned microclimate (See Figure 11.1) and use standard soil ecology and microbiology protocols (Hall 1996; Bramble 1995; and LaMotte 1999) to determine the densities of various critical microbe and invertebrate populations as well as the levels of key inorganic nutrients (See Figures 11.2 and 11.3). They use this annual baseline (together with data from previous years) to perform a statistical analysis, looking for any anomalous findings (see Figure 11.4, p. 176), and then return to the field to search for potential causal explanations for a specific anomaly of interest to them.

Each team next posits a hypothesis based on their findings and prepares to test it with a controlled field experiment of their own design. One year, for example, a group found that there were substantially fewer protozoa in our wetland microclimate than expected for such a damp environment (see Figures 11.5 and 11.6, p. 176). However, this site was receiving significantly more sunlight that year due to a downed tree, and so the girls hypothesized that something about the increased exposure to the sun was causing the lower protozoa density. After a literature search, they learned that UV radiation has been shown to cause protozoa to descend deeper into the soil column, and consequently, the students predicted that the protozoa in the anomalous site must be farther down in the ground. The team designed and set up an experiment to block the UV light from the sun incrementally (see Figure 11.7), took samples from different depths of soil (Figure 11.8), and eventually supported their hypothesis (E.S.S.R.E. 2004). Similar investigations are central to the E.S.S.R.E. program, and the findings of the interns become a permanent part of the program database. All interns finish their research by learning how to write formal papers of the kind professional scientists would submit for peer-reviewed publication (and which the program posts on its website as a teaching resource for others), and when appropriate, the girls make presentations to the school administration and the Board of Trustees to advise them about any significant changes in our Forest Conservation Easement that need addressing.

The last thing the groups do is to adapt the experimental protocol they used in their research into a more general method for investigating a topic such as theirs in other places and schools. Each team designs and creates a website with instructions on how to do an experiment similar to their own (see Figure 11.9), and they include on these sites resources for learning about and studying the environment in a wide range of physical and socioeconomic situations using alternative, low-cost materials and low-tech methods. These sample lessons then become part of program's permanent and ongoing outreach efforts, and through workshops and the E.S.S.R.E. home page, the girls and their work continue to help equip others with better learning tools for becoming more knowledgeable and aware of local environments and to engage and manage those environments more rationally and effectively.

Figure 11.1

Figure 11.2

Figure 11.3

Figure 11.4

Figure 11.5

Figure 11.6

Alignment With *More Emphasis*

One of the central challenges the National Science Education Standards make is to change how teachers and students interact in the learning process itself. With a greater emphasis on guiding, sharing, and supporting rather than telling, dispensing, and enforcing (NRC 1996), the Standards encourage classroom teachers to see themselves as collaborators in the learning process rather than imperious experts and to approach their task in ways that place the student rather than the teacher at the center of the learning process. In this regard, E.S.S.R.E. is a model program. Previous interns have described it in letters of support for grant funding as one that

> embodies [the] teaching ideals of creativity, innovation, and hands-on experience for [the] students and creates a path by which young students, specifically young women, can experience interactive science education and gain early independent research experience.... [E.S.S.R.E.] provided guidance and support in helping [us] formulate unique research questions, yet allowed independence and creativity. Working in groups, we struggled to narrow the research focus, utilize previous scientific data, and devise a working hypothesis. Through this process, we learned important lessons of teamwork and compromise, as well as leadership and communication skills. I remember the excitement of devising our own research questions, and the ability to uncover previously unknown knowledge. [The program] made it a priority to cultivate this excitement and creativity among us.... [It] has influenced my life and educational path in profound ways.

Other girls have commented on their end-of-program evaluations that

> [the] desire to work *with* us on our research rather than above us shows a unique and innovative approach to teaching [and that] rather than treating the summer program interns as mere "workers" [the teaching staff] imbued us with a desire to push ourselves and accepted the challenge of becoming our tutors and instructors, not just our "bosses." Therefore... although interns worked from nine to three–what would normally seem like a fairly long day for high school girls in the summer–the time flew by and many afternoons we would have to be gently nudged out of the classroom, still chuckling at the soft humor of [a] last remark....

E.S.S.R.E. is obviously "providing opportunities for scientific discussion and debate," "sharing responsibility for learning," and "supporting a classroom community with cooperation, shared responsibility, and respect" (NRC 1996), and thus, it has demonstrated for nearly a decade now how the interface of cooperative student and teacher research teams can develop and implement new ways of teaching and learning that are more effective. Moreover, through our web lessons and workshops (see Figure 11.10), the interns, teaching assistants, and project director have helped educators around the country to do likewise.

Yet we have strived to do more than simply model good teaching with E.S.S.R.E.; we have sought to model good content as well. In the work the young women do, the emphasis is placed on well-crafted scientific arguments and a quick look at their work reveals just how sophisticated student learning can be when provided with both the opportunity and the expectations. For instance, one of the things the

Figure 11.7

Figure 11.8

Figure 11.9

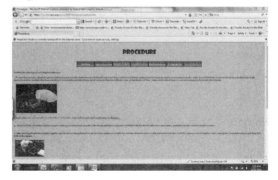

Figure 11.10

interns must learn is to place their specific research question within the larger context of scientific understanding. Thus, when one group noticed in 2007 an unusual pattern to the distribution of iron in one of our microclimates, they searched the necessary literature to discover that

> iron is a very significant and essential element in the environment and is identifiable in the soil by its rusty red or orange color (O'Reilly 2002).... Its ferrous iron form is essential to plant growth because it is essential in the production of chlorophyll, which in turn is significant because it is necessary for photosynthesis... However, the most plentiful form of iron in the soil is ferric iron, and ferric iron compounds cannot be absorbed by the roots. The pH of soil determines which form of iron is present, with ferrous iron available in acidic soil and ferric iron available in basic soils. Hence as the soil's pH rises, plants must struggle and generally do not succeed in using the iron as effectively....

This group went on to learn that "one key factor that influences soil pH, and consequently iron levels is the process of leaching" and

> further research led us to the theory that it was the water from a surrounding stream that washed the excess iron into the site. We decided to test this hypothesis by deliberately leaching soil plots where the correct pH and iron relationship was being observed....

Such reasoning is precisely what the content standards are challenging science educators to achieve, and it is clear that these young women left E.S.S.R.E. knowing how to "[communicate] science explanations" and understanding "science as argument and explanation" (NRC 1996).

E.S.S.R.E. interns also leave knowing how to manage an immense amount of ideas and information. The girls must collect and analyze over 800 individual pieces of data over the course of the three weeks (see Figure 11.11) and that is just to complete the biota survey and not their group research investigations! They genuinely get to experience science at its most elemental, and as a consequence, they discover for themselves what all good science education strives for: the validity and reality of their own unique "scientific" voices. As Tom Peri, science department chair at another area school has written to grant agencies,

> It seems very appropriate to include some of [my students'] reactions to the ESSRE in their own words. "We like how we learned about environmental problems concerning soil. We

Figure 11.11. 2007 Site 4 Summary

		$\bar{y}$	Mode	Median	s	s2	Min.	Max.	Range	n	s/√n
Chemical Tests											
Salts	Ca (ppm)	895.45	1400	1000	476.16	226728.35	150	1400	1250	12	137.46
	K (ppm)	46.67	0	55	37.71	1422.04	0	90	90	12	10.89
	Cl (ppm)	83.33	50	75	44.38	1969.58	50	200	150	12	12.81
	Fe (ppm)	20.63	25	20	5.9	34.81	7.5	25	17.5	12	1.7
	Al (ppm)	15.42	7.5	7.5	21.07	443.94	0	80	80	12	1.76
	Mg (ppm)	12.92	10	10	7.53	56.7	5	25	20	12	2.17
	Mn (ppm)	7.125	0	0	9.85	97.02	0	25	25	12	2.84
Inorganic Nutrients	P (ppm)	91.67	100	100	12.31	151.54	75	100	25	12	3.55
	Nitrite (ppm)	0	0	0	0	0	0	0	0	12	NA
	Nitrate (ppm)	11.04	7.5	7.5	5.98	35.76	5	25	20	12	1.73
	Ammonia (ppm)	1.88	0	1.25	2.85	8.12	0	10	10	12	0.82
Soil pH		7	7	7	0.17	0.03	6.8	7.4	0.6	12	0.05
Microbes											
Heterotrophic Bacteria	(#/cc soil)	3.65E+05	150,000	3.75E+05	2.29E+05	5.26E+10	8.00E+04	7.00E+05	6.20E+05	12	66220.2
Fungi	Yeast (#/cc soil)	5.72E+04	8000	3.90E+04	5.80E+04	3.36E+09	6.00E+03	2.00E+05	1.94E+05	12	16740.3
	Molds (#/cc soil)	3350	100	7.50E+02	6.23E+03	3.88E+07	1.00E+02	2.10E+04	2.09E+04	12	1797.91
	Combined (#/cc soil)	5.89E+04	all	4.06E+04	5.73E+04	3.28E+09	8.10E+03	2.01E+05	1.93E+05	12	16527.6
Algae	(# / mm2)	2.58	all	1.02	4.85	23.52	0.19	17.57	17.38	12	6.79
Protozoa	(#/g soil)	83592.83	all	67893.5	64760.4	4.19E+09	9639	194696	185057	12	18686.1

saw what it was like to be a scientist," wrote a 2007 participant. A 2005 participant said, "Chemical testing was fun mixing everything together and learning what levels are normal for an environment. It was an awesome experience." "I really felt like a scientist and I like how we took our data and findings to the web," said a student from 2006. All of these girls have the common experience of doing actual science in a meaningful way in a real life situation [and] that kind of experience provides a wonderful context for their course content and a strong foundation for further study in high school and beyond.

Students Who Engage: A Case Study

As of July 2010, over 120 young women from a variety of schools in the Baltimore area (six public, four parochial, and five private) have successfully completed and statistically verified 10 biota surveys and 30 field studies. Their results have helped guide our school's Strategic Plan—introducing artificial turf fields to stop the disruption of the nitrogen cycle in the woods that the girls have documented—and their numerous web lessons have resulted in the publication of a lab manual for high school teachers co-authored by three of the interns (Brock et al. 2008). The program continues to receive vigorous support from the Toshiba America Foundation, Cary Institute of Ecosystem Studies, Waksman Foundation for Microbiology, Paul F-Brandwein Institute, National Science Foundation (through the PAEMST program), and Human Capital Development, Inc. (a local corporate sponsor), and we now actually fund approximately one-third of the program through the royalties earned off the teaching kits the girls have developed (in addition to the lab manual) which Flinn Scientific sells.

Furthermore, while there are no formal requirements for admittance to the program other than a genuine interest in science (and the teacher recommendation to back it up), participants in E.S.S.R.E. have used it as a stepping stone to far more competitive programs, including the University of Maryland's Center of Marine Biotechnology, the Johns Hopkins University's Space Telescope Science Institute, the VA Palo Alto Health Care System Spinal Cord Injury, the University of Melbourne (Australia) Faculty of Land and Food Resources, and Protein Design Labs.

Over 50% of those who have now graduated high school have gone on to pursue postgraduate training in the sciences and engineering, and two of our interns received full-expense, four-year science scholarships to attend their higher education institutions. Of the 41 young women who have now actually graduated college, 14 have chosen to enter STEM fields (3 MD/PhD, 2 Biomedical Engineering, 2 Neuroscience, 2 Science Education, 1 Aerospace Engineering, 1 Industrial Mechanical Engineering, 1 Environmental Engineering, 1 Biology PhD, and 1 Math/Information Sciences), and our total participants included 16% Asian-Pacific Islanders, 15% African-Americans, and 7% Hispanics. Indeed, E.S.S.R.E. has been so successful at accomplishing its goals that we started partnering in 2007 with the Growing Girls and Gardens Program of the Baltimore City Public Schools to provide graduates of this low-income middle school program guaranteed slots in our program in order to help support their long-term success in the sciences, and so far, 100% of these girls have continued to take rigorous science courses every year of their high school careers.

But the preceding information was just "the facts," and while it is obviously possible to read into them evidence of the *More Emphasis* recommendations of the NSES, the "numbers" do not tell the entire tale. E.S.S.R.E.'s real triumph is in the transformed lives of the young women themselves, and to illustrate, I would like to take a closer look at the experience of two recent interns.

The story begins, oddly enough, in our school's driveway. Preliminary research done in the 9th-grade soil ecology program (Brock 2005) had revealed that car exhaust from the carpool line appeared to be having a negative impact on the density of soil protozoa in the front lawn of the school. My students had hypothesized that the sulfur dioxide from the fumes was leaching into the soil, where it was altering the pH in much the way acid rain does by interacting with moisture to form hydrogen sulfate. Furthermore, their investigation suggested that, in fact, there was a correlation between the key environmental indicators of pH, sulfate, and protozoa density. However, the nature of their project was not such that they could verify or substantiate their findings.

Enter two of my summer interns, MariaLisa and Julianne. In their first year in the program, both young women had developed an intense interest in the impact of runoff on soil health. MariaLisa had completed a detailed analysis of the how fertilizer from campus was altering various critical components of the nitrogen cycle in our woodland research sites, while Julianne had worked on the team that investigated leaching. Both girls, therefore, had the necessary aptitude and attitude to tackle a problem such as the one involving the car exhaust, and when they returned the following summer as E.S.S.R.E. teaching assistants, they decided (in addition to their numerous other duties!) to tackle the issue my 9th graders had raised.

There, after an extensive controlled experiment (much of which was conducted on their own time), they not only substantiated the earlier findings; they also determined that school land-scaping decisions were actually playing a role in the situation. As they stated in their presentation prepared for the Trustee's Building and Grounds Subcommittee,

> This past spring the freshmen completed a scientific investigation of the front lawn. Their findings suggested that as the distance away from the carpool line increased, the levels of sulfuric acids decreased and the pH levels increased; these were the expected trends.... Our results very closely parallel those obtained in the spring. As distance increased away from the carpool lane, levels of pH increased and the levels of sulfate decreased.... [However] our data shows that the [plant] buffer [covering half of the research site] helps to stabilize the environment of the front lawn soil. The small difference in protozoa density [in correlation to distance] on the side which has the buffer is insignificant because of the stable environ-ment the buffer creates; while in the plots without the buffer protozoa levels correlate the way we'd expect: as the distance from the carpool lane increased, the protozoa level also increased (this follows the idea that protozoa would prefer living in less acidic soil, which would be found farther from sulfur dioxide emissions).... .

The girls then went on to argue that

> the real issue is the side of the lawn without a buffer, where there are extreme differences in the amounts of protozoa [in correlation to distance], proving that the plant buffer is a useful

tool in keeping a neutral lawn. Because of this, we suggest that a row of bushes be incorporated into the landscaping that wraps behind the memorial garden. Not only would it be visually pleasing, but it would also contribute to Roland Park's environmental sustainability….

Even if the story ended here, it is a clear example of students successfully "[engaged] intelligently in public discourse and debate about matters of scientific and technological concern" (NRC 1996). But the story continues. Following graduation MariaLisa continued keeping track of the situation through regular e-mail communication and Julianne worked with our new Environmental Education and Sustainability Coordinator to promote awareness of the situation. A new landscaping plan was, in fact, developed and presented to the Trustees for final approval. Thus, not only is E.S.S.R.E. meeting a local need for supporting the interest of young women in learning about and pursuing possible careers in science and meeting the larger need of improving environmental education; it is showing young women that their "voice" genuinely matters. As a teacher dedicated to the education of young women, could I ask for anything more?

Conclusion

The Environmental Science Sumer Research Experience for Young Women has now played a positive role in the lives of young women from Baltimore for nine years. Providing the target audience the rare opportunity as high school students to work on the frontier of a relatively nascent scientific field, the program has received national recognition with a SeaWorld/Busch Gardens/Fujifilm Environmental Excellence Award in 2006 and was a finalist for Cable's Leaders in Learning Award in 2009. Indeed, E.S.S.R.E. has been so successful in its efforts to promote environmental education and young women's interest in science that our Board of Trustees has given approval to seek to endow the program permanently, and it has had so positive an impact on our community that we have already raised 50% of the necessary finances to do so. E.S.S.R.E. is changing lives for the better in Baltimore and beyond, and we plan to do so for many years to come.

Acknowledgments

Special thanks must go to Peter Groffman of the Cary Institute of Ecosystem Studies for being our consulting expert in all things soil all these years; to my first teaching assistants, Beccy Josowitz and Becky Polan, for single-handedly creating the infrastructure for this program; and to Jean Waller Brune, Head of School, for her unwavering support of a dream that started with 17 girls doing field studies out of her garage. Without each of their unique contributions, E.S.S.R.E. wouldn't be what it is today.

References

American Association of University Women (AAUW). 2004. *Under the microscope: A decade of gender equity projects in the science*. Washington, DC: American Association of University Women Educational Foundation.

Bramble, J. E. 1995. *Field methods in ecological investigation for secondary science teachers*. St. Louis: Missouri Botanical Garden.

Brock, D. 2005. It is the "little things" that can change the way you teach. In *Exemplary science in grades 9–12: Standards-based success stories*, ed. R. Yager, 1–9. Arlington: NSTA Press.

Brock, D., K. Brockmeyer, K. Loya, and M. Torres. 2008. *Soil ecology lab manual*. Batavia, IL: Flinn Scientific.

Committee on Science, Engineering, and Public Policy. 2007. *Beyond bias and barriers: Fulfilling the potential of women in academic science and engineering*. Washington, DC: National Academies Press.

Cothron, J. H., R. N. Giese, and R. J. Rezba. 2000. *Students and research: Practical strategies for science classrooms and competitions*. 3rd ed. Dubuque, IA: Kendall/Hunt.

Environmental Science Summer Research Experience (ESSRE). 2004. ESSRE Research findings and results. Available online at *https://faculty.rpcs.org/essre/ ESSRE%20Research.htm*

Hall, G. S., ed. 1996. *Methods for the examination of organismal diversity in soils and sediments*. Paris: CAB International.

LaMotte. 1999. *Model STH series combination soil outfit instruction manual*. Chestertown, MD: LaMotte Company.

National Research Council (NRC). 1996. National science education standards. Washington, DC: National Academies Press.

National Zoological Park. 2009. Smithsonian institution's monitoring and assessment of biodiversity program research protocols. Available online at *http://nationalzoo.si.edu/ConservationandScience/MAB/research/protocols.cfm*

O'Reilly, S. 2002. Iron-rich soil can help remove lead; Manganese also important. *Virginia Technological University Science Blog*. Available online at *www.scienceblog.com/community/older/2002/A/20027034.html*

Samuels, M. L. 1989. *Statistics for the life sciences*. Englewood Cliffs: Prentice Hall.

The CHANCE Program

Transitioning From Simple Inquiry-Based Learning to Professional Science Practice

Jacqueline S. McLaughlin and Kathleen A. Fadigan
Pennsylvania State University

Setting

Two settings build the premise for this chapter and comprise the Connecting Humans and Nature through Conservation Experiences (CHANCE) program. The first setting is a summer field course in Costa Rica and the second is the regular school classroom during the academic year. The first setting builds the foundation for the second.

Introduction

Imagine it is summer time in Costa Rica. Picture a group of approximately 20 preservice and inservice high school science teachers along with biology undergraduate students assisting local and international nongovernmental organizations (NGOs) with an assortment of research projects and activities that take place directly within this tropical environment. The entire group actively participates in conservation efforts in Gandoca, near the border with Panama on the Caribbean coast, to protect the Leatherback sea turtle by assisting researchers and staff of Widecast in guarding nests and hatchlings from predators (including humans).

At the Organization of Tropical Studies' La Selva Biological Station in Puerto Viejo, the group interacts with several researchers from all across the globe. This site is located at the confluence of two major rivers in the Caribbean lowland of northern Costa Rica and spans 1,600 hectares (3,900 acres) of tropical wet forests and disturbed lands. During their stay here the group breaks into small teams and assists some of the researchers with their data collection. For example, one team helped to clear coffee plant saplings from square meter plots to enable researchers to study the ways in which rain forest land recovers after having been cleared for agricultural purposes. Another team ventured into the rain forest to monitor army ant colony behavior dynamics. Another team worked in an on-site laboratory to analyze the antipredatory defense mechanisms of *Dendrobates pumilio*, the deadly strawberry dart frog.

Another leg of the journey transports the group to the John H. Phipps Biological Station of the Caribbean Conservation Corporation in Tortuguero. Here the group assists with

Figure 12.1. The CHANCE Logo

recording the nesting activity of the green turtle, *Chelonia mydas,* including measuring and tagging the female and monitoring nesting coordinates as they relate to beach dynamics, especially erosion. The teachers and biology students engage in discussions with scientists who are studying hatchling viability, life history of juvenile turtles in and around convergence zones of the open ocean, levels of phthalate pollutants in embryonic yolk, and telemetry data from long distance adult migrants.

Now, fast-forward to the fall. The new school year is in session. Teachers who participated in the summer field course in Costa Rica have successfully completed a follow-up workshop where they learned how to implement the CHANCE interactive multimedia research modules into their science teaching (*www.chance.psu.edu*). These freely available, web-based multimedia tools are designed to engage and empower both teachers and students with authentic research data from scientists who are currently investigating a wide variety of environmental issues in the life sciences.

Confident in their own knowledge of conducting field research in conservation biology, the teachers (CHANCE Fellows) are prepared to share their personal field experiences and expertise with their students, and to use technology equipped with higher-order inquiry-based learning tools to do so. Throughout the academic year these teachers are integrating the CHANCE modules into their curriculum where they best fit. While doing so, the teachers are actively engaging their students in lively discussions and debate based upon the authentic research data provided by the CHANCE modules that their students are manipulating and interpreting. The modules are encouraging students to go beyond the core content found in textbooks, and to consider the implications of the research data they are analyzing. With the help of the modules, the teachers are encouraging their students to look at both local and global big pictures.

Introduction to the Program

Initiated in 2004, the CHANCE program is a partnership between The Pennsylvania State University (PSU) and the Pennsylvania Department of Education (PDE) and has had several funders since it originated. It was developed to (a) engage Pennsylvania inservice and preservice high school teachers in environmental science and conservation biology through the excitement of hands-on field research in selected ecosystems in Costa Rica and (b) enhance the way Pennsylvania teachers teach and the way Pennsylvania students learn by using technology to bring real-world scientific data into the classroom.

The overarching goal of the CHANCE program is to produce levels of understanding, knowledge retention, and transfer that are greater than those resulting from traditional lecture-based classes by blending teaching and basic research. These goals are achieved by implementation of the field course in Costa Rica, development and implementation of the CHANCE modules, and professional development.

The CHANCE modules are a set of internet-based, environmentally themed learning tools that use genuine research data (McLaughlin 2006, 2010a, 2010b). Targeted toward high school science students, each module features a student-as-researcher approach through student manipulation of a data set contributed by scientists who are currently investigating the topic. The CHANCE modules attempt to capture the true nature of scientific research by providing students opportunities to work with scientists and their research data. These tools are also filling the niche in the transition from simple inquiry-based learning (textbooks, less student responsibility) to professional science practice (research, more student responsibility), while at the same time promoting environmental literacy and stewardship.

Thus far, nearly 100 preservice and inservice secondary science teachers from across the nation have been trained in the field in Costa Rica. Although the field course immerses teachers in scientific research and discourse at the global level, it is not imperative for teachers to participate in the field course in order to successfully utilize the CHANCE modules. The CHANCE director conducts three or four training sessions per year at local and national venues. Over 500 teachers have participated in such workshops. These workshops provide teachers with training to begin implementing the modules in the classroom.

Because all CHANCE modules are free to any educator with internet access, students and teachers in classrooms anywhere in the world who have not completed the CHANCE field course or a module training workshop can still implement the modules. Teachers with a strong research background can independently complete a module from start to finish in a few hours and be ready to effectively use it with their students. Note that a "teacher guidelines" link is provided for each module to assist any teacher, experienced or novice, with pedagogical tips and content delivery. Additionally, module users are encouraged to write to the module authors, scientists, and CHANCE director with questions concerning a specific module's use or content.

Methodology

The Modules: Structure and Development

Each CHANCE module is a web-based, multimedia science lesson that guides students through a series of web pages (somewhat similar to a web quest) that strategically builds upon basic environmental science concepts aligned to both Pennsylvania and national standards. Using multimedia technology, the CHANCE modules transport the students away from textbooks to a virtual research location. The format is similar to the idea of a virtual field trip, but instead of being guided by a teacher, students are chaperoned by a renowned research scientist. After completing several interactive and challenging assignments that cover essential core concepts, students are taken to a research location where they engage in the same process of scientific inquiry used by the researcher who co-authored the module. Students then manipulate authentic data in order to draw conclusions about an environmental issue. For example, the amphibian module, "Amphibians as Indicators of Environmental Change," starts with some simple taxonomy, graduates to an international mapping activity and the life cycle of Costa Rica's

Golden Toad (*Bufo periglenes*), then progresses to an activity that details the science behind the global greenhouse effect. Students then analyze a published research study by Dr. Rick Relyea, associate professor and director of the Pymatuning Laboratory of Ecology at the University of Pittsburgh, that provides evidence that specific amphibian populations in western Pennsylvania are declining due to pollution from pesticides.

To date, seven modules have been developed and are already in use in high schools and educational organizations throughout the world. Completed CHANCE modules focus on the following global topics: invasive plant species; raptor migration; amphibians as indicators of environmental change; sea turtle nesting behaviors and survival; deciduous forest biodiversity; species extinction; and global climate change (Figure 12.2). Modules in the planning stages include topics on life and evolution, watershed restoration, waste disposal, water pollution, and burning culm—the main waste product of coal incineration—to produce electricity.

In addition to the CHANCE curriculum activity pages, each module's home page contains pertinent background information, teacher guidelines, Pennsylvania state standards addressed by the module, suggested websites for further exploration, and related classroom activities. The teacher guidelines strongly recommend that each teacher complete a selected module in its entirety before students interface with this technology. Key is that teachers (think teachers as researchers) master the material and the process themselves and become empowered to be facilitators of learning and work to create interactive classrooms wherein students take on more responsibility. It is also recommended that teachers take the time to provide an overview of the objectives and content of each activity page with their students in order to orient them and help them to navigate the site. For example, the activity on page three of the most recently developed module, "Global Warming: Turning Up the Heat," a short introduction to isotopes and the methods used in this module to study surface temperature from ice layers is highly suggested. Each activity

Figure 12.2. CHANCE Modules Table of Contents

page of the modules takes 30–45 minutes to complete. (Each module consists of five or six activity pages.) Thus, teachers may elect to have students only complete certain activity pages based upon what standards and topics they are planning to cover and the time they have to devote to teaching specific topics.

In the module "Global Warming: Turning Up the Heat," students are transported to Costa Rica to work with Dr. Deborah Clark, an ecologist at the Organization of Tropical Studies, Costa Rica, known for her research on carbon stores and fluxes in tropical forests. At the module's home page, the user is presented with background information on the layers of the Earth with a specific focus on the biosphere. At the end of this introduction students are presented with an interactive activity that demonstrates the layers of both the Earth and the atmosphere on a global scale. Students are prompted to record their responses to nine open-ended questions in CHANCE's unique assessment recording tool, the Progressive Notebook.

The Progressive Notebook is structured so that students type their responses in a text box located below the question. Once students have finished typing their responses and click on the "next question" button, the module records and saves their responses. At any point in the module users can click on the Progressive Notebook button to view their work. The Progressive Notebook page resembles a spiral-bound research notebook and each question and answer is presented in the order it was completed. The page can be printed or saved as a PDF file for assessment or study purposes.

After completing their responses, students can then move on to the next activity (designated as pages) by clicking on a link at the bottom of the page or by using the sidebar menu. The next activity of this module features the carbon cycle and how humans are shifting carbon from the lithosphere to the atmosphere. After a brief introduction the viewer is prompted to complete an activity that is part of the Environmental Protection Agency's *Climate Change Kids* website *(http://epa.gov/climatechange/kids/carbon_cycle_version2.html)*. The third activity introduces users to measuring atmospheric variations in CO_2 levels through graphing data collected in Mauna Loa, Hawaii by the Scripps Institute of Oceanography since 1958 (Keeling's curve). In the fourth activity, users travel to Greenland to work with Dr. Richard Alley, Penn State University paleo-climatologist, and take ice core samples to look at surface temperatures dating back thousands of years to understand past climate fluctuations. Users are personally introduced to Dr. Alley and his research by viewing a video produced specially for the module.

Students take actual measurements of canopy tree annual growth over a 16-year period in the fifth activity, gathering published data to demonstrate how increased canopy temperature affects rain forest productivity (Figure 12.3). In the final activity users meet Dr. James Hansen, Director of NASA's Goddard Institute for Space Studies, to learn about the urgency that lies within our present climate crisis, and the reality of the gap that exists between scientists and policy makers. Users also learn how and why the ice is melting in the Arctic Circle, and how and why we must act now to lower CO_2 emissions into the atmosphere.

Each CHANCE module is developed through an intricate collaboration between teachers, university experts in teaching and learning, a prominent scientist, the program director, research institutes, governmental organizations, nongovernmental organizations, and instructional technology experts. This partnership has helped the final products to be a valid representation of

Figure 12.3. Activity 5 of CHANCE Module "Global Warming: Turning Up the Heat"

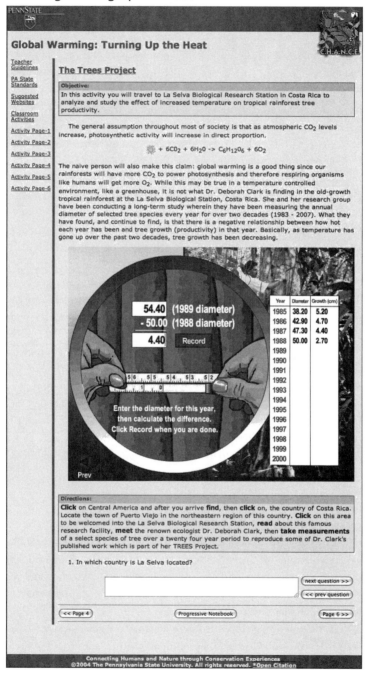

current scientific research, address the most recent pedagogical practices in science education, and utilize effective and engaging instructional technology from animations, videos, and virtual explorations. These factors stimulate students to explore, critically think about, and understand key environmental science issues and biological concepts.

New module development is an intensive process that incorporates the feedback and expertise of all of the stakeholders. Each year the CHANCE Director selects a few teachers who participated in the CHANCE field course in Costa Rica to co-develop new modules. This process, known as storyboarding, includes writing a detailed, accurate description of the content for each and every page of a new module. The module development team works together to craft the text, activities, links, and animations.

The CHANCE director trains the teachers and scientists on storyboarding, and mentors them throughout the development of each new module outline. Ten preservice and inservice teachers have participated in this process since 2004. The storyboarding training lasts for at least one month, and a storyboard takes approximately three months to complete.

This carefully constructed professional development opportunity allows preservice and inservice science teachers to focus on how people learn science, using theories of multimedia learning, such as techniques to prevent cognitive overload (Mayer and Moreno 2003) and how to assess and address student preconceptions. Knowledge-centered presentations of materials must consist of both a conceptual framework and a set of coherent facts. CHANCE participants also learn how to integrate formative assessment that allows learners to interact with new knowledge on-screen and on paper for a balanced curricular approach (Bransford et al. 1999; Donovan and Bransford 2005; Duschl et al. 2007).

Teachers are carefully paired with experts in the selected field of study (called CHANCE mentors) to acquire current and relevant information and guidance. Storyboards are continuously edited and co-authored by the director. Modules are extensively reviewed for scientific accuracy by CHANCE mentors and other research scientists. When finalized, the new module is sent to the CHANCE instructional design staff for electronic page development and uploading; this process is also supervised by the director and the selected teacher authors. Finally, high school teachers from around the nation field-test the new module. From start to finish each module takes six to eight months to complete.

Results: Assessment of Learning

Various research and evaluation measures have been implemented to evaluate and improve both the field portion of the program and the modules.

The Field Course

The CHANCE field course strictly employs the "Field Course Experiential Learning Model" (Zervanos and McLaughlin 2003; McLaughlin 2005), which evolved from repeated short-term study abroad (embedded) field course experiences in selected biomes around the world over a six-year period. Assessment of student learning guided the development of the integrated three-part course model including pre-trip assignments, a field-based trip experience, and post-trip assignments that encourage the integration and application of what has been learned (McLaughlin and Johnson 2006).

One goal is to move students beyond simple memorization of facts to higher-order domains of application and integration (Bloom et al. 1956; Anderson and Krathwohl 2001). For this reason, the course has embedded performance assessments that require students to perform authentic tasks to demonstrate knowledge and skills. Students are expected to demonstrate the ability to apply conceptual understanding. Students keep daily journals similar to field journals used by scientists to organize and document field observations. This authentic assessment replicates an activity in which scientists engage. The journal activity encourages and provides evidence of higher-order thinking. Journal assessment is supplemented by a post-trip assignment in which students provide open-ended responses to questions. This assignment helps students connect readings and the field experience and provides an opportunity to demonstrate their level of learning.

A "Student Assessment of Learning Gains" was also developed. This instrument asks students to rate and verbally elaborate on the extent to which specific learning activities contributed to specific knowledge and skill outcomes. For example, students are asked to rate and discuss the extent to which the field experience contributed to their understanding of basic principles of biodiversity, to their understanding of the relationship between human activity and the environment, and other general biological concepts. Similar questions explore how the experience contributed to gains in scientific skills and higher-order cognitive gains. Assessment findings from three cohorts of students and preservice teachers (62 participants total) participating in the field course showed cognitive and affective learning gains of students in all learning activities evaluated (McLaughlin and Johnson 2006). Additionally, data revealed that problem-centered field experiences enhance the student's knowledge of biodiversity, increase their understanding of how humans impact ecosystems, and influence how they would make behavioral decisions relative to biodiversity in the future.

A follow-up study is currently planned to longitudinally track former CHANCE field course participants to assess ways the program has influenced their teaching. Teacher participants have provided informal reflective feedback in numerous follow-up conversations. The quotes below are representative.

> First, the experience of field work in Costa Rica has helped invigorate me in the classroom; second, the wealth of experiences, pictures, knowledge, and research picked up during the trip (and in the accompanying assignments) has provided me with tons of new material for my classroom; and third, the available modules are helping me integrate interactive technology into my classes as well as being valuable curricular material in their own right. *Teacher participant, 2007.*

> When I went to Costa Rica as part of the CHANCE program, I had just graduated college and was about to embark on my first year of teaching 8th grade science. Looking back, my experiences doing the field research with Costa Rica's rich beauty... truly molded my teaching philosophy. The wonderment and excitement I felt made me want to recreate that for my students. If I could evoke these same kinds of emotions and inspiration in my

students, they would learn. Bringing biodiversity and inquiry to my classroom through the CHANCE program has been the most influential part of my teaching career. *Teacher participant, 2004.*

The CHANCE field course changed my life. Since that experience, I've become more passionate and enthusiastic about teaching biology. It is incredible how much I can now impact and influence my students about some serious ecological issues simply by sharing pictures and stories about my trip to Costa Rica. *Teacher participant, 2005.*

I am a high school Biology and AP Environmental Science teacher. The CHANCE program was the BEST professional development program that I have participated in, in over 20 years of teaching. The interactive approach to Conservation Biology, integrating research was invaluable. I learned the importance of providing my high school students with this skill, and the modules make this task truly possible. In addition, seeing, feeling, touching, participating in programs in Costa Rica was absolutely a life changing experience that I bring to my classroom every day. *Teacher participant, 2007.*

The Modules

Recent research supports the effectiveness of the CHANCE modules over more traditional instruction methods (McLaughlin 2010a, 2010b). This research compared six high school classes in three different Pennsylvania high schools using traditional textbook lecture and recitation methods to six classes using the CHANCE modules to teach a biology unit on ecology (covering similar core concepts and standards). Figure 12.4 shows the test results for the textbook and module groups (n = 212, with 115 students total in the textbook groups and 97 students total in the module groups); Table 12.1 shows the mean, standard deviation, statistical significance, and confidence levels for the test results shown in Figure 12.4. The results in Figure 12.4 and Table 12.1 reveal a statistically significant difference between the pre- and posttest scores of the two groups. Both groups were tested before beginning study of the scheduled ecology unit; following study of the ecology unit topics covered in chapters 4, 5, and 6 in the course textbook; and three months after the ecology unit concluded.

Data analysis revealed that between the pre- and posttests, the student participants who used the CHANCE modules as the primary mode of learning instead of the textbook scored higher on concept assessments given for the content in each chapter. Posttest scores for the module group showed a gain of 12.99 percentage points, while the textbook group showed a gain of 5.91 percentage points. A t-test evaluating the difference in the final posttest scores between the textbook and module groups indicated statistical significance at the p = 0.001 level (Table 12.1 and Figure 12.4). Importantly, both student populations showed no significant difference at the pre-assessment stage (the groups varied by only 1.91 percentage points), suggesting that both groups overall possessed similar academic ability.

Figure 12.4. Textbook vs. Module Groups Test Data (All Groups)

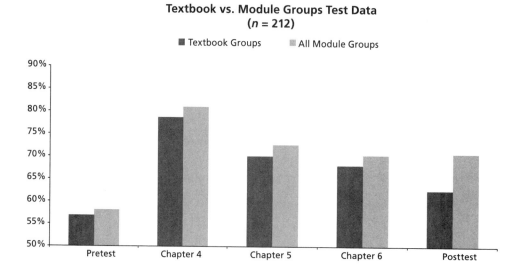

Textbook vs. Module Groups Test Data
(*n* = 212)

■ Textbook Groups   ■ All Module Groups

Table 12.1. Mean, Standard Deviation, Statistical Significance, Confidence for High School Group Tests

	Number of Students	Mean Grade	Standard Deviation	t	p-value
All Textbook Groups Pretest	115	55.792	4.048		
All Module Groups Pretest	97	57.708	3.638	0.896	0.186
All Textbook Groups Posttest	115	63.107	3.666		
All Module Groups Posttest	97	69.937	4.222	3.114	0.001

Overall, student achievement has risen when the curriculum is supplemented with the CHANCE modules. It is important to note, however, that when the overall data (*n* = 212) is separated into groups depicting academic levels of the students involved, analysis reveals that the Gifted High Potential (*n* = 61) and College Prep (*n* = 92) groups of students who utilized the CHANCE research modules scored higher on the three chapter Concept Assessments than the students using only their textbook, and that the Applied Group (*n* = 59) faired lower (Figures 12.5–12.7; Table 12.2). This result is not surprising since these students may not have had a strong interest in science in the first place, especially being classified in a "lower" group designation. The results stress an urgent need for relevant technology-based curriculum materials for noncollege bound students as well as those with an academic career path.

Table 12.2 shows the average test scores as per academic level for the test results plotted in Figures 12.5–12.7. The results revealed a statistically significant difference between the pre- and posttest scores of the Gifted High Potential (p = 0.012 level) and College Prep groups (p = o.004 level). For the Applied Group, however, results did not reveal a statistically significant difference between, the pre- and posttest scores (p = 0.37 level).

Figure 12.5. Textbook vs. Module Groups Test Data (Gifted High Potential)

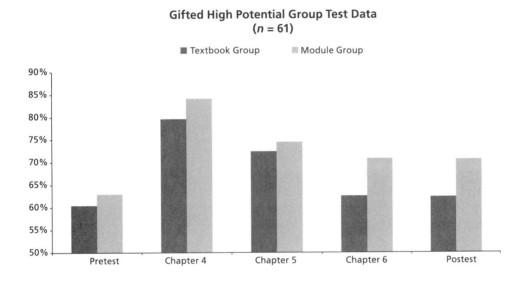

Gifted High Potential Group Test Data
(n = 61)

Figure 12.6. Textbook vs. Module Groups Test Data (College Prep Group)

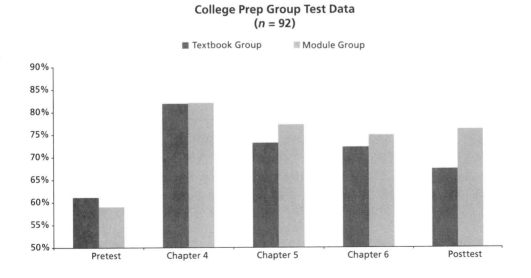

College Prep Group Test Data
(n = 92)

Figure 12.7. Textbook vs. Module Groups Test Data (Applied Group)

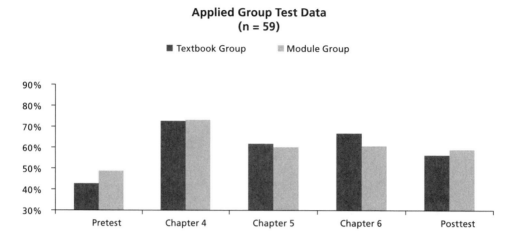

Table 12.2. Average Test Scores as Per Academic Level

	Pre	Ch 4	Ch 5	Ch 6	Post
Gifted High Potential (*n* = 61)					
Textbook Group	60.44%	79.62%	72.40%	62.55%	62.36%
Module Group	63.00%	84.14%	74.48%	70.81%	70.64%
College Prep (*n* = 92)					
Textbook Group	61.36%	82.46%	73.63%	72.66%	67.55%
Module Group	59.24%	82.69%	77.76%	75.54%	76.65%
Applied (*n* = 59)					
Textbook Group	42.80%	72.70%	61.80%	66.77%	56.36%
Module Group	48.76%	73.18%	60.16%	60.74%	59.04%

We have only measured learning outcomes data for the CHANCE modules for a short time, so our next research project will involve a longitudinal study to determine whether the stronger learning outcomes among high school students are attributable to the CHANCE modules and not to the novelty effect (Bauer 2004). New research modules specifically designed for undergraduate college students are presently under development, and a plan is underway to assess learning outcomes of this group in the near future.

All groups of high school students involved in this study also completed pre- and postattitudinal surveys to assess their attitudes toward learning science, doing research, and using technology. Overall, student data reveal that today's youth have positive attitudes toward learning about science and understanding how research is conducted. The surveys also assess students' perceptions of textbooks. In postsurvey responses to the statement "I find my biology textbook interesting," 52 % of the students disagreed. Yet, in response to the statement "I feel comfortable using computer-based resources as learning tools," 81% of the students agreed. This information stresses the notion that this generation of youth considers technology part of their daily life. Figure 12.8 shows student responses to the survey questions posed following conclusion of the ecology unit part of the course. The survey used a Likert scale ranging from 1 = Strongly Disagree to 5 = Strongly Agree.

Figure 12.8. Post-Study Survey Responses (from top to bottom)

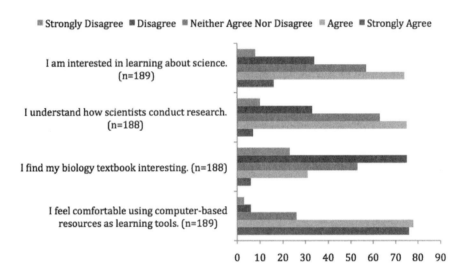

Additional project data provides a qualitative look at students' and teachers' perceptions of the use of the modules, feedback on the CHANCE module design, appropriate high school level science content, and the likelihood of teacher adoption (McLaughlin and Arbeider 2008). Following a two-hour workshop on CHANCE module use in the classroom, 68 teachers (10 preservice and 58 inservice teachers) evaluated the "Invasive Plant Species in Pennsylvania" module using both the cyberguide for Website Design, and the cyberguide for Content Evaluation (MacLachlan 1996). To gauge the likelihood of adoption, a subset of 46 teachers, including 10 preservice teachers, also completed a brief implementation survey noting their likelihood of using the module in the future. Responses to the cyberguide evaluations were equally strong among both preservice and inservice teachers. Teachers' responses served as a feedback mechanism for module design. Of major interest was that the preservice teachers showed more interest and likelihood to implement digital technology in general than their more experienced counterparts. In free response comments, preservice teachers even noted that they would attempt to implement the CHANCE modules during their student teaching practicum. Preservice teachers commented:

> The modules really are engaging. I felt an adrenaline rush as a teacher as I worked through the module. All I could think of was that my students will love this and actually learn something meaningful about the environment as they use it.

> I liked the animations, pictures, activities, and research scenarios. I can't wait to try the other modules.

Based on these preliminary findings, a follow-up study that includes demographic information and pedagogical inventory data will investigate these differences more thoroughly.

The CHANCE director has conducted numerous full- and half-day workshops over the past several years at national, state, and local teacher conferences, as well as in schools and other professional development settings. Sessions were designed to help teachers become comfortable with the technological format of the CHANCE modules and employ strategies to best incorporate the material into their existing science curriculum.

Workshop participants completed evaluation forms at each session. These forms provide a wide range of feedback from teachers. After completing a three-hour CHANCE module training session, teachers frequently comment on their increased comfort with using research to enhance the learning of science concepts and their plans to integrate higher-level scientific inquiry into their teaching. Teachers expressed intentions to use the CHANCE modules and incorporate more opportunities for scientific research through current articles, guest speakers, and experiments.

In spring 2009 a national textbook company began conducting blind reviews of the modules. The company recruited a cohort of teachers from across the nation to respond to a series of questions about the organization, level, depth of content, and functionality of the modules, including an in-depth review of the global warming module. Now complete, the analysis has been overwhelmingly positive. One of the most frequently stated weaknesses is the time involved. Many reviewers expressed concerns about their time constraints in the classroom.

Some of the reviewers' comments regarding the global warming module include:

> I feel this is very effective at allowing the students to teach themselves the topic of global warming. I love the students being involved in the research and graphing the data. Activities #1 and #2 are useful review of the core science.

> I think the chance to see ice cores and tree diameters will really help the students. The people who introduced their research and the student manipulation are wonderful. I would definitely have my students do those two activities but may not always have them do the module from start to finish (time constraints).

> I definitely would assign this module. I would take my students to the computer lab and we would complete each segment together and stop for sharing thoughts and concerns. This allows me to supplement if I feel it necessary.

Conclusion: Beyond the Modules

Beyond simply using the modules as a teaching tool, teachers and students are discussing and debating environmental issues. They are working together, looking at the larger community—including businesses, schools, and government—and establishing partnerships.

For instance, in the winter of 2009 the CHANCE program with support from the pharmaceutical company, sanofi-pasteur, and an international roofing company, ATAS International, Inc., developed the CHANCE 2009 Music and the Environment contest. The idea was to have students watch the five-star rated YouTube video of Dr. Richard Alley singing a song on global warming to the tune of "Proud Mary" (*http://chance.psu.edu/idol.htm*). Pennsylvania high school students worked in groups of three to five to revise the lyrics of one of their favorite songs to stress the importance of an environmental issue. Students performed their songs live in a school setting, recorded, and uploaded them to YouTube. A group of judges selected the winning entry. The top winners received a cash prize, which must be used in their school to implement an environmentally focused project that allows their school to become more sustainable. The winning video is also showcased on the CHANCE website. The 2009 winners will be using their $2,500 prize to create a community garden and an outdoor theater on a parcel of land on the East Pocono High School campus in Swiftwater, Pennsylvania. The judges agreed with the winning students that this garden/theatre will give their community a place to gather for special events, such as an Earth Day celebration, and will also nurture the environment.

Other teacher participants have started environmentally themed student organizations. Two teacher participants have led separate projects to plan and implement a school community garden for aesthetics and outdoor class activities. Other teachers have organized local field experiences for their students—one teacher now takes her high school students to Costa Rica—and have become more involved in science fair projects. Many teachers are continuing their professional development by membership in professional organizations, completing additional workshops and training, and connecting with scientists. Many teachers have instigated a "current events" activity in their weekly lesson plan that focuses on local, national, and international environmental issues.

Alignment With National Science Education Standards: *More Emphasis*

On numerous levels the CHANCE program heeds the call for science education reform set forth by the NSES. On the teaching front the CHANCE program offers teachers an opportunity to take charge of their curriculum by providing tools that can be used in whole or in part to assist with building scientific inquiry skills while using authentic research data as a vehicle for learning. Each module sets the stage and provides high-quality fuel for lively discussion about scientific topics that address local and global concerns.

In terms of professional development, the CHANCE program offers rich opportunities for learning science through investigation and inquiry. Through the CHANCE field course, teachers directly contribute to research studies with a global impact. Teachers are frequently asked to compare and contrast their home environment and community with the ones that they visit in Central America. Upon their return home several teachers continue to pursue scientific inquiries through new partnerships with scientists who assist them with the development of new CHANCE modules. CHANCE participants often keep in touch with their colleagues, providing support for each other and in some instances partnering on new educational endeavors.

Each CHANCE module places a heavy emphasis on students' development of inquiry abilities. The environmental themes allow for easy integration of other subject matter disciplines. In a time when many budgets do not allow for many laboratory or field inquiries, the CHANCE modules offer an affordable way for students to investigate science and develop their own explanations for the results they produce. The modules promote such science-process skills as communicating and analyzing data, and making scientific arguments.

School science programs can benefit from the CHANCE modules not only because they are freely available online and support the state and national standards, but also because they address issues that are of increasing interest to today's youth. Further, the modules reduce dependence on textbooks and lecture, encouraging a more student-centered approach to learning.

Conclusion

Both elements of the CHANCE program, the field course and the modules, offer teachers an up-to-date alternative to traditional professional development and curriculum opportunities. Based upon ongoing research and evaluation findings, the CHANCE program continues to evolve. This program supports the learning of scientific inquiry skills and the vision of the NSES.

The CHANCE research modules take full advantage of the digital aptitude of students and new teachers and promote inquiry-based learning by requiring students to explore, observe, question, hypothesize, manipulate, analyze, and think critically about real science data and information from accredited research programs in Pennsylvania and around the world. Experienced teachers embrace the authentic science that forms the basis for CHANCE modules and pick up easily on the interactive nature of the materials. This computer-based interactive approach is particularly attractive to secondary students, most of whom enjoy the virtual experiences today's technology affords. Best of all, CHANCE modules promote active learning by providing opportunities for students to participate as individuals directing their own learning process.

Indeed, Robert DeHaan reports in his exemplary review on the state of science education in the United States that, "using real research strategies to teach has profound effects on student learning, and could have profound effects in promoting a scientifically literate society and a reinvigorated research enterprise" (2005). The CHANCE team believes that only a well-informed citizenry will be qualified to create policies and develop programs that will help stem the extensive alteration of our natural world. Publishing data in a peer-reviewed article is important; yet only a select few can actually read and understand targeted and complex data. Conveying complex data to the level that allows high school students to think and ponder real-world questions is one of the most worthwhile endeavors that scientists can undertake; for it is in the imperative of the messages, delivered to those who will build on what they learn in their classrooms, that our very futures depend.

The Pennsylvania Department of Education recommends the CHANCE field course and research modules as ways of helping high school teachers and their students meet the nine state standards in environmental science and ecology. Because most states must meet similar standards, the CHANCE program provides a viable framework for improving or reforming high school biology education nationwide.

Authors' Notes

- The CHANCE modules are available for free at *http//:chance.psu.edu/toc.htm*
- To apply to participate in a field course in Costa Rica, attend an upcoming workshop on module training, or learn about other outreach initiatives like the CHANCE Idol Video Contest visit *www.chance.psu.edu* or e-mail Jacqueline McLaughlin, founding director, at JShea@psu.edu.
- CHANCE is the winner of the 2005 "Bringing the World to Pennsylvania" Award from the Pennsylvania Council on International Education (PaCIE).

Acknowledgments

For belief in CHANCE and its impact, we thank Patricia Vathis (PDE), Ellyn Schlinder (sanofi pasteur), and James Bus (ATAS International, Inc.), and all other partners, sponsors, and collaborators for their endorsement and backing (complete list on our website). We are especially grateful to the following NGOs for their devotion to the CHANCE field course and allowing us to blend real-world research with science education: The Organization of Tropical Studies (OTS) in La Selva, Costa Rica; The Smithsonian Tropical Research Institute in Bocas del Toro, Panama; Widecast (Duke University) in Gandoca, Costa Rica; and, The Caribbean Conservation Corporation, Costa Rica. For the amazing assistance with module development and design, Dr. McLaughlin is beholden to all the CHANCE high school teacher authors, research scientist mentors, and multimedia design specialists who used their talents to help create a different type of pedagogical tool. We are also appreciative of the expert assistance of Ronald Heft in preparing the figures and tables showcased herein. Finally, we want to thank the Pennsylvania State University for their continual support of the CHANCE program and our personal growth as scholars in the arena of engaged scholarship.

References

Anderson, L. W., and D. R. Krathwohl, eds. 2001. *A taxonomy for learning, teaching, and assessing: A revision of bloom's taxonomy of educational objectives.* New York: Addison Wesley Longman.

Bauer, K. 2004. Conducting longitudinal studies. *New Directions for Institutional Research* 121: 75–90.

Bloom, B. S., M. D. Engelhart, E. J. Furst, W. H. Hill, and D. R. Krathwohl. 1956. *Taxonomy of educational objectives: Handbook 1: Cognitive domain.* New York: David McKay.

Bransford, J., A. Brown, and R. Cocking, eds. 1999. *How people learn: Brain, mind, experience and school.* Washington, DC: National Academies Press.

DeHaan, R. 2005. The impending revolution in undergraduate science education. *Journal of Science Education and Technology* 14 (2): 253–269.

Donovan, M., and J. Bransford. 2005. *How students learn: Science in the classroom.* Washington, DC: National Academies Press.

Duschl, R. H. Schweingruber, and A. Shouse. 2007. *Taking science to school: Learning and teaching science in grades K–8.* Washington, DC: National Academies Press.

Johnson, D. K., and J. L. Ratcliff. 2004. Creating coherence: The unfinished agenda. *New Directions for Higher Education* 125: 85–96.

MacLachlan, K. 1996. *WWW CyberGuides.* Available online at *www.cyberbee.com/guides_sites.html*

Mayer, R. E., and R. Moreno. 2003. Nine ways to reduce cognitive load in multimedia learning. *Educational Psychologist* 38 (1): 43–52.

McLaughlin, J. S. 2005. Classrooms without walls: A banana plantation, a turtle nest, and the random fallen tree. *International Educator* 14 (1): 52–54.

McLaughlin, J. S. 2006. Real world biology education blends teaching and basic research. *The Science Teacher* 73 (8): 48–53.

McLaughlin, J. S. 2010a. Reimagining science education and pedagogical tools: Blending research with teaching. *EDUCAUSE Quarterly* 33 (1). *www.educause.edu/eq*

McLaughlin, J. S. 2010b. A multimedia learning tool that allows high school teachers and their students to engage in scientific research. In *Technology leadership in teacher education: Integrated solutions and experiences*, eds. J. Yamamoto, C. Penny, J. Leight, and S. Winterton, 25N35. Hershey, PA: IGI Global.

McLaughlin, J. S., and D. Arbeider. 2008. Evaluating multimedia-learning tools based on authentic research data that teach biology concepts and environmental stewardship. *Contemporary Issues in Technology and Teacher Education* 8 (1): 1–19.

McLaughlin, J. S., D. A. Arbeider, J. L. Stine, T. R. Dugan, and P. J. Besong. 2010. *CHANCE: Connecting humans and nature through conservation experiences.* Available online at *http://chance.psu.edu*

McLaughlin, J. S., and D. K. Johnson. 2006. Assessing the field course experiential learning model: Transforming collegiate short-term study abroad experiences into rich learning environments. *Frontiers: The Interdisciplinary Journal of Study Abroad* 13: 65–85.

Zervanos, S. M., and J. S. McLaughlin. 2003. Teaching biodiversity and evolution through travel course experiences. *The American Biology Teacher* 65: 683–688.

Engaging Students in Content Learning and Scientific Critique Through a Nanoscience Context

Clara Cahill and Minyoung Song
University of Michigan

Cesar Delgado
University of Texas at Austin

Setting

Nanoscience is increasingly visible in scientific endeavors, new technologies, and engineered products. Current understanding about nanoscience among the American public is minimal, so perceived risks and benefits are assumed on the basis of personal belief or perception rather than scientific understanding (Batt, Waldron, and Broadwater 2008; Cobb and Macoubrie 2004). Conceptual understanding of nanoscale science is essential for the public to have a balanced and informed perspective on new nanotechnologies, understand the benefits and potential hazards of these emergent technologies, and thus be able to make informed decisions concerning this new and increasingly prolific area of science and technology (Roco and Bainbridge 2006). Citizens and students must develop a sense of what constitutes scientific evidence of the positive and negative effects as well as side effects of nanotechnology on human health and wellbeing, the environment, and technological products. As such, nanotechnology provides a rich and rigorous context for the development of scientific literacy, one that is especially exciting because it can offer insights into the nature of science as new discoveries are made. In this chapter, we present a curriculum focused on helping middle school students build conceptual understanding of topics essential to nanoscale science through engagement in the evaluation of scientific claims.

Introduction

The curriculum was piloted during a two-week summer science camp for middle school students in a diverse midwestern school district. Throughout the camp, students were asked to evaluate the claims of advertisements and product manufacturers, and to make decisions and critique the uses of nanotechnology in society. Two complementary instructional strands focused on concepts essential to size and scale and size-dependent properties, two of the "Big Ideas" of nanoscale science (Stevens et al. 2009). Instruction and student investigation in each strand revolved around a central driving question, which focused student investigations and helped to drive student learning (Krajcik and Czerniak 2007). Students investigated size and scale through the driving question of "How can nanotechnology keep me from getting sick?" The size-dependent properties strand focused on one of two related driving questions: "How can nanotechnology help treat lung cancer?" and "How can nanotechnology help treat asthma?" These questions enabled students to critically evaluate claims made about the potential benefits and uses of nanotechnology in consumer and medical products, provided an opportunity for students to develop and investigate their own questions related to nanotechnology, and created rich instructional and motivational contexts for exploring essential scientific content (Blumenfeld, Kempler, and Krajcik 2006). The construction and identification of strong, well-supported scientific explanations, including claims, evidence, and reasoning (McNeill et al. 2006), were highlighted throughout the two-week camp, across the curricular strands, to promote the development of well-supported, scientifically based strategies for evaluating the advertiser's claims and making decisions related to nanotechnology.

The overall curriculum was inquiry-based, and developed using the Construct-Centered Design (CCD) process to coordinate learning goals, curriculum, and assessment in a principled manner (Pellegrino et al. 2008). CCD combines learning-goals-driven instructional design (Krajcik, McNeill, and Reiser 2008), backwards design (Wiggins and McTighe1988), and evidence-centered design of assessments (Mislevy et al. 2003) to facilitate the coordination of the entire learning environment around helping students achieve the target learning performances (Pellegrino et al. 2008). CCD involves selecting and elaborating a "big idea" or instructional standard in order to specify and characterize particular concepts on which to base the instructional intervention. After the big idea is fully elaborated, specific learning goals related to the content are defined. At this stage, student prior knowledge, common non-normative ideas, potential cognitive or conceptual difficulties related to the unpacked learning goals, phenomena illustrative of the learning goals, and required and expected prior knowledge are considered in conjunction with the student grade level, the length of the instructional intervention, and the overall goals of the instructional setting. This enables the definition of targeted *claims,* which are the skills, knowledge, and competencies the designer wishes to claim the students will demonstrate at the end of the instructional intervention. Next, the *evidence* is defined. Evidence is defined in the form of a learning performance or cognitive task that specifies how the targeted knowledge and skills might be manifested in student performances, behaviors, or artifacts in such a way that the claim can be evaluated. Finally, assessment *tasks* are developed to elicit student performances that will generate the evidence (Pellegrino et al. 2008). Learning activities are then developed around a "driving question" or rich instructional context, with the targeted claims, evidence, and tasks driving the learning of students (Krajcik and Czerniak 2007). Thus, the CCD process promotes the principled development of instructional materials, and ensures strict focus on helping students construct the targeted knowledge and skills within contextualized instruction.

The Curriculum

Strand 1 (Size-Dependent Properties): How Can Nanotechnology Help Treat Asthma and Lung Cancer?

In this strand, students were asked to investigate how asthma or lung cancer medication can be made more effective by nanosizing the pieces of medicine in dry inhaler spray, focusing on the question about whether reducing the size of solid pieces could actually improve their rate of dissolution. This series of lessons connects primarily to chemistry topics, in particular connecting surface and size to overall rates of dissolution. To that end, this strand was constructed around four interconnected claims generated from the CCD process that describe what we expected students to be able to do with their developing understanding of these concepts. See Table 13.1.

Table 13.1. Claims, Evidence, and Tasks for Size-Dependent Properties Strand

Claims	Evidence	Example of a Task
The student is able to use mechanistic reasoning to explain what happens when a solid dissolves in a liquid, emphasizing that dissolution occurs at the surface of solids.	Mechanistic reasoning is used to diagram and/or explain the process of dissolution. The mechanistic reasoning may include a description of the macroscopic dissolution behavior, particles of solute and solvent, agency of the solvent in dissolving the solute, spatial association between the solute and solvent, and a description of the process of dissolution.	Task 1: Students draw a diagram explaining what happens when something dissolves from the perspective of being "Zoom[ed] in all the way, as far as you can go, so that you can show what is really happening."
The student is able to explain that changing the size and shape of materials impacts rates of dissolving, and to apply these principles to increase rates of dissolution.	Prediction using observation-based principles or mechanistic reasoning that, if two samples of a material have the same mass but different size or shape, their rate of dissolution could be different.	Task 2: Students write a scientific explanation of why smaller grains of solute dissolve more quickly than the same mass of larger grains of the same solute, and describe how this principle can be used to increase rates of dissolution.
The student is able to exemplify and explain the importance of object surface as the interface between two materials using observation-based principles and/or a particle-based mechanistic explanation.	Application of the principle that rates of dissolution are related to surface area to predict and explain the role of surface area in determining rates of dissolution when comparing dissolution rates between two objects made of the same material.	Task 3: Students make a prediction supported by an explanation, about whether two different objects of identical mass but different shapes will dissolve at different rates.
The student is able to identify and explain how changing the size or shape of materials impacts the number of exposed solid particles in the material.	Explanation that materials broken up into smaller pieces have higher surface area, and, therefore, more solid particles exposed to the solvent.	In task 2 and task 3, student use of a particle model of solutes that demonstrates how a change in size or shape relates to the number of exposed particles provides the desired evidence.

This lesson sequence is designed to help students understand the particulate nature of matter in solids and liquids, and the connection of the particulate nature of matter to the properties and behavior of materials. There are six 2.25-hour lessons in this sequence:

Lesson 1: Nanotechnology and Treating Asthma

Students are introduced to a claim made by fictional drug manufacturers that their 200 nm-sized dry-powder asthma medication works faster and more effectively than micron-sized dry powder medications currently available (See Figure 13.1). Although the advertisement is fictional, the context is real: both asthma and lung cancer medications are currently being tested for their effectiveness (Dames et al. 2007; Sham et al. 2004; Sung, Pulliam, and Edwards, 2007). Students are asked to identify claims that the manufacturer is making. They critically evaluate these claims, developing questions that provide the context for the subsequent lessons.

Lesson 2: How Fast Can It Dissolve?

Students design and conduct an initial experiment to determine whether breaking up a material into smaller pieces makes it dissolve faster. Student groups create a poster explaining the experiments they conducted, detailing the claim they were making relative to the medication context, and providing reasoning and evidence they gathered to support their claims.

Lesson 3: Concentrating on Treating Asthma

Students further investigate the claims of the manufacturer about the effectiveness of the drug, using probeware and visual observation to compare the rate of increase in the concentration of a "drug" as it dissolves, when it is in large pieces as compared to when it is in small pieces.

Lesson 4: What Happens When Things Dissolve?

Students next look more closely at the process of dissolution to better understand why nano-sized formulations of medication dissolve more quickly than micron-size formulations. Students watch the process of dissolution for table salt through the microscope at 40x, 100x, and 1000x magnification to gain an understanding of how dissolution progresses, and of the importance of surfaces during the process of dissolution. With the magnification, they can see that a tiny grain of salt dissolves from the outside, much like an ice cube melts from the outside. Students observe simulations and models of solids, liquids, and solids dissolving in liquids, and build their own models in order to better understand the process of dissolution and the connection between dissolution and concentration. Emphasizing the connection between dissolution and concentration enables students to use the particle model to explain rates of concentration increase, expanding their understanding of and experience with the applications of the particle model.

Lesson 5: On the Surface

Students connect what they have learned about dissolution to volume and surface area, through observing how vinegar visibly diffuses into agar imbued with acid/base indicator, and connecting these observations directly to models of solids and liquids. Students compete to design ways to shape the agar in order to maximize the overall diffusion and minimize the amount of time it takes for the acid to diffuse to the center of the agar. Groups orally present explanations justifying their design decisions. Returning to the initial context, students are asked to consider other ways to make dry asthma or lung cancer medication that will dissolve rapidly. Students design lego-block models to maximize dissolution rate, and are asked to provide evidence and reasoning supporting the idea that their designs will maximize the rate of dissolution.

Lesson 6: Making Decisions and Recommendations

In the final lesson, students use what they have learned to create a critical evaluation of the nano-formulation of the asthma or lung cancer medication. Students present their findings, make recommendations for further study, and explain any questions or concerns that they have regarding the asthma or lung cancer treatments.

Figure 13.1. Nanopran advertisement: Contextualizing Advertisement for Fictitious Asthma Medicine Used in Size-Dependent Properties Strand

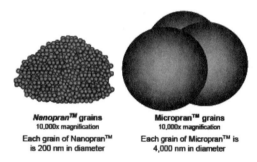

Strand 2 (Size and Scale): How Can Nanotechnology Keep Me From Getting Sick?

This second instructional strand focused on size and scale and was contextualized via the driving question, "How can nanotechnology keep me from getting sick?" and focused on bacteria and viruses (Delgado, Short, and Krajcik 2009). Students learned about a person who died after contracting methicillin-resistant *Staphylococcus aureus,* a drug-resistant bacterium. Then students were asked to evaluate whether to purchase toilets incorporating a nanotechnology that produces ultra smooth finishes, claimed to reduce the buildup of germs. This context led to an examination of the size of surface features of the toilet's finish and of bacteria and viruses. The overall learning goal for the camp, which was unpacked and elaborated using the CCD process, was:

> Students will know that there are unseen worlds that are too small to be seen with the naked eye. These include the micrometer, nanometer, and subnanometer worlds. Students will know of several submacroscopic* objects, including the atom and the cell, and have an idea of their relative and absolute sizes. [* submacroscopic: objects too small to be seen with the unaided eye]

To achieve this learning goal, students engaged in a series of five 135-minute lessons designed to support their understanding of the relative and absolute sizes of submacroscopic objects. Students used a projecting microscope to trace the width of a hair and to determine how many hairs would fit across a millimeter and then used this information to calculate the average width of a hair. Then, students increased the magnification of the microscope to enable them to see bacteria and skin cells as well as a hair, and determined the size of these objects by calculating how many bacteria or skin cells would fit across a hair. This same process was used to enable students to determine the size of viruses, atoms, and other submacroscopic objects using a computer simulation designed especially for the camp, based on the idea of lining up smaller objects across the surface of objects of known size in order to determine relative and absolute sizes. A second computer simulation enabled the students to compare the sizes of all of the submacroscopic objects to the head of a pin, by counting and observing aurally and visually the length of time it would take to line up each object across the pinhead, if placed on the pin at a constant rate (Song and Quintana 2009). These and other activities enabled the students to develop a sense of the relative size of some submacroscopic objects, and to use this understanding to create advertisements for the nanosurfaces using explanations and models of why reducing surface features to nanoscale sizes could help reduce germ buildup (Delgado et al. 2009).

Methodology

Participants and Methods of Assessment

The summer camp enrolled 32 middle-school students from a diverse midwestern school district, in which 56% of the students qualify for free or reduced-price lunch. Students were divided into sections based on grade level. Each section had similar instructional activities and assessments, contextualized with different driving questions.

Student learning was assessed in multiple ways, before, during, and immediately following instruction. Student content understanding in the size and scale strand was assessed by pre- and postinstruction interviews. In the size-dependent properties strand, student learning was assessed at multiple points through the curriculum, through written formative assessments, pre- and postwritten assessments, and summative performance assessments designed to scaffold understanding and promote sense-making, while probing for student explanatory levels of understanding (Fellows 1996; Rivard and Straw 2000; Ruiz-Primo et al. 2004). In addition, a comparative retention assessment of explanation-embedded content was conducted six months after the completion of the summer camp. This assessment compared students who applied to and attended the summer science camp (n = 19, 62.5% of attendees), and students who applied but chose not to attend the summer science camp (n = 18, 62.1% of nonattendees). The retention assessment was used in combination with the pre- and post-camp assessments to evaluate the effectiveness of the instructional intervention for helping students develop normative, causal-mechanistic explanations of dissolving and the relationship between size, shape, and rates of dissolution. These groups are similar on the basis of age, race, gender, school, and quality of application essay, and we posit that the nonattending students' responses are similar to what the attending student responses would be if they had not participated in the summer camp. Thus, the comparative assessment enabled us to evaluate the robustness of any changes in student explanations over time. The comparative retention assessment, in contrast to the pre- postinstruction assessment, utilized primarily multiple-choice items, with three short-answer questions and 14 multiple-choice questions. The multiple-choice questions had multiple correct responses, which corresponded to different levels of explanations of dissolution and the relationship between size, shape, and rates of dissolution. As a result, the retention assessment enables us to gauge students' ability to choose between different types of explanations. Because of the cognitive differences in the process of choosing a best response in a multiple-choice test, or crafting a complete response in an open-ended question (Ward, Frederiksen, and Carlson 1980), these different assessments enable us to evaluate how summer camp attendees construct their own explanations, and how their ability to choose well-supported scientific explanations differs from similar students who did not attend the camp. Student demographics are detailed in Table 13.2.

Table 13.2. Student Demographics

		Total Attendees	Comparison Attendees for 6-month retention study	Comparison Nonattendees for 6-month retention study
Gender	Female	16	10	8
	Male	16	9	10
Last Grade completed	6th	16	12	9
	7th	12	6	8
	8th	4	1	1
Race	Non-Caucasian (African American, Hispanic, biracial, or multiracial)	21	13	13
	Caucasian non-Hispanic	9	4	5
	No response	2	2	0
Overall		32	19	18

Results and Evidence of Successes

To provide a deep sense of student learning outcomes, we will focus our discussion of the student outcomes on the size-dependent properties strand. However, students experienced significant gains in the size and scale strand as well. Student overall factual knowledge increased with an effect size of 0.83 ($p<0.001$ for paired sample t-test comparing means). Summer camp participants reached a level similar to that of high school students who had not experienced this curriculum (Delgado et al. 2009).

In the size-dependent strand, students improved significantly from the pre- to the postinstruction assessment, and demonstrated more sophisticated understanding of appropriate scientific explanations of dissolving and rates of dissolution than the comparison group six months after the camp. Figure 13.2 indicates the percentage of students whose work demonstrated at least the minimum evidence for each claim. Students improved significantly in their abilities to explain the process of a solute dissolving in water (Claim 1), (effect size (ES) = 1.23, $p<0.001$ for paired two-sample t-test). Student explanations and understanding of the relationship between size and shape and the rate of dissolution of materials (Claims 2–4) also increased significantly (ES = 1.63, $p<0.001$ for a paired two-sample t-test). The comparative retention evaluation suggests that camp attendees selected more sophisticated scientific reasoning than nonattendees to explain the process of dissolution (Claim 1) (ES = 1.77, $p<0.001$ for independent sample t-test), and to explain the relationship between size and shape and the rate of dissolution of materials (Claims 2–4)(ES = 1.94, $p<0.001$ for independent sample t-test).

Figure 13.2. Percentage of Students Achieving Each Claim. Pre- and Postinstruction Comparison Showing Statistically Significant Mean Student Learning for All Four Claims in the Size-Dependent Strand

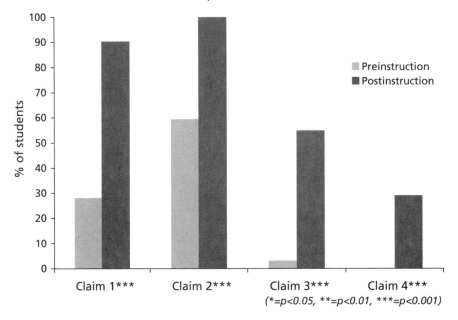

($*=p<0.05$, $**=p<0.01$, $***=p<0.001$)

Overall, the pre-post analysis indicated that student-generated responses to general questions about what happens when solutes dissolve tended to become more causal-mechanistic as a result of instruction (ES = 1.05, $p<0.001$, for a paired two-sample t-test), indicating that students developed the skills to identify and use more scientific types of explanations. Students used descriptive responses more frequently prior to instruction, indicating a reliance on observation rather than theory or mechanism in their interpretations of phenomena. For example, one sixth-grade female student responded to a request to explain what really happens that causes things to dissolve by stating solutes "get smaller," depicting bubbles emanating from the solute as it dissolved. In contrast, after instruction, the same student represented water molecules "carrying away" particles of solute as the solute dissolved. Another sixth-grade female student responded to the same question by describing how "bubbles are rising and popping… the water is getting mistey [sic] with the [solute]," a macroscopic description of dissolution. In contrast, after instruction, the same student described and drew how "H_2O is taking particles [of solid] away, increasing the concentration." These changes indicated that, as students' conceptual understanding improved, their reasoning and explanation strategies improved as well.

In addition to changing the ways students described how things dissolve when directly asked to explain dissolution, our qualitative and quantitative evaluations suggest that this curriculum was successful in changing the reasoning strategies and explanatory frameworks students used to evaluate claims about the capabilities of nanotechnologies and the promises of nanoscience specifically (See Figure 13.3, p. 213, for a graph of the changes in the types of explanations and reasoning students provided when asked to evaluate advertisers' claims about nano-size medications). Thus, students developed skills for making and identifying scientific explanations; and developed the aforementioned content understanding related to dissolving. As mentioned earlier, one of our primary objectives in the summer camp was to help students develop the ability to use scientific reasoning to evaluate new nanotechnologies and make decisions, which supports student achievement of National Science Education Standards Goal 3 by encouraging students to approach the evaluation of advertisers' claims from a scientific perspective (NRC 1996).

Initially, half of the students used nonscientific reasoning, such as belief-based justifications, non-normative reasoning, or simply did not provide any justification when they evaluated claims about nano-size grains of medicine. For example, "Allan," a seventh-grade male student believed at the beginning of instruction that the advertised claims that nano-size medicine would dissolve faster than micron-size medicine "probably isn't accurate because most advertiser [sic] aren't." This critical stance indicates no use of scientific reasoning to evaluate the claims. "Harriet," a 13 year-old female initially responded to the advertised claims by stating "I think [the advertiser's claims] are true, because I think nano is the future." This initial claim is uncritical of the claim, and relies on personal belief rather than scientific reasoning.

Forty-three percent of students initially applied principle-based reasoning to help evaluate the products. We define principle-based reasoning as a type of reasoning using observation-based principles to support a claim. For example, "Clarise," a 13-year-old female, believed the advertiser's claims "because it is a finer powder it will take less time to dissolve," while "Kathy," a 13-year-old female stated that she believed the advertiser's claims because "It's faster and more efficiently because it's smaller… and they said it was improved." This type of reasoning employs the implicit or explicit observation-based principle that smaller things dissolve faster to evaluate the advertiser's claims. Although many material behaviors and characteristics may seem dissimilar and unrelated when only the directly observed effect is considered, the underlying structures and mechanisms causing behaviors may be similar. Thinking about phenomena from a mechanistic perspective can help highlight these similarities, and can contribute to deep, unified understanding of the way things behave and operate (Glennan 2002; Salmon 1984). Mechanistic reasoning often includes both the principle and a causal explanation of the principle. No students used mechanistic reasoning to help explain or evaluate the claims initially.

In contrast, 96.9% students provided some type of legitimate scientific reasoning after instruction, suggesting that the instructional intervention may have helped to increase the use of scientific reasoning in evaluating specific products and making decisions. By the end of the instruction, "Allan" used mechanistic reasoning to evaluate the claim, stating that he believed that "Nanodox [nano-size lung cancer medicine] works better because it's smaller. If you had three small spheres and a big one, the three smalls would dissolve first because smaller stuff dissolves at a faster rate. With several small things they can all be worked by the water at once so they dissolve faster. [Also] smaller things have fewer particles each so there's less to dissolve. So I believe that Nanodox will work better because of the evidence I have gathered." However, he still had concerns about the medicine, asking, "Will the Nanodox hurt the person because it kills all rapidly producing cells not just cancer? Will it get all the cancer? Is it more expensive?" This indicates that the instructional intervention helped the student increase the scientific reasoning used in his decision-making about this aspect of nanoscale materials.

At the end of instruction, "Harriet" also used mechanistic reasoning in her evaluation, stating that she believed that "Since Nanopran [nano-size asthma medicine] is small it dissolves faster, much faster than Micropran [micron-size asthma medicine] (we've tested and experimented this to see if this is true and it is)…Since Nanopran dissolves faster, it gets to the problem faster. With Micropran, it's small but…the water can't get to the particles inside [of the grain] as fast so it takes more time to work." In addition, Harriet expressed some critical questions about the use of this new technology, worrying whether it could "go places it shouldn't because its so small." Her responses at the end of instruction demonstrated enhanced understanding as well as improved scientific reasoning. Kathy's responses also indicated basic mechanistic reasoning after instruction, stating that "Nanopran works faster than Micropran because it's smaller and it's easier to dissolve. When things are smaller, they're more vulnerable…and it's easier for the water molecules to attack…" Kathy's response demonstrates that she has developed the ability to use a particle-based mechanism to explain how things dissolve and why different sizes of the same material dissolve at different rates.

The comparative retention analysis suggests that attendees' use of scientifically based forms of reasoning was significantly greater than nonattendees. Figure 13.3 shows the average number of students who used each type of response in the pre-post performance assessments, and Figure 13.4 shows the average percent of students' overall responses in the retention study for each type of response.

Figure 13.3. Types of Reasoning Students Used in Evaluation of Advertisers' Claims During the Pre-Post Instruction Assessment. (* $p<0.05$, ** $p<0.01$, *** = $p<0.001$, significance calculated using Wilcoxon signed-ranks test)

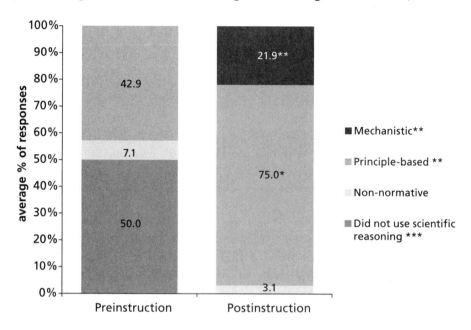

Figure 13.4. Types of Reasoning Students Used in Evaluation of Claims During the Retention Assessment (* $p<0.05$, ** $p<0.01$, *** $p<0.001$, significance calculated using Wilcoxon signed-ranks test)

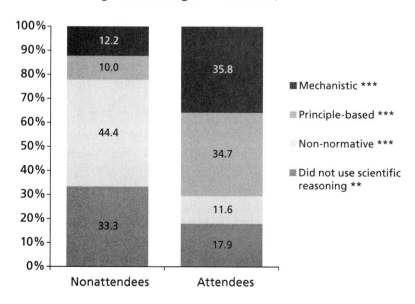

Conclusion

Our curriculum focused on helping students build conceptual understanding of topics essential to nanoscale science, and build competence in making decisions and evaluating claims about nanotechnology. Through critically exploring the causal mechanisms enabling the unique functionalities behind two specific nanotechnologies—nanosmooth surfaces and nanosized particles of medication—we helped the students develop and improve their abilities to evaluate nanotechnologies from a scientific perspective. Students' skills and content understanding improved during the curriculum as they engaged in multiple iterations of critique, exploration, and development of scientific reasoning and explanations. Students were motivated and engaged by the process of analyzing claims made by the advertisements and one another. Providing the opportunity for students to make and defend their own claims, as well as to challenge and appraise the claims made by other students, helped them understand the foundational importance of well-documented and presented scientific evidence and reasoning.

Acknowledgments

This program was supported by the Ypsilanti Public School District; the Health Occupations Partners in Education Program at the University of Michigan; the University of Michigan School of Education; and the National Center for the Learning and Teaching of Nanoscale Science and Engineering. We would like express our sincere thanks to all of our instructors and staff for the 2008 summer camp, including Ashima Mathur, Yolanda Campbell, Charmica Abinojar, Paula Sizemore, Lucie Howell, Kenneth Tang, and Thomas Ambrose.

References

Batt, C. A., A. M. Waldron, and N. Broadwater. 2008. Numbers, scale, and symbols: The public understanding of nanotechnology. *Journal of Nanoparticle Resaerch* 10 (7): 1141–1148.

Blumenfeld, P. C., T. M. Kempler, and J. S. Krajcik. 2006. Motivation and cognitive engagement in learning environments. In *The Cambridge handbook of the learning sciences,* ed. R. K. Sawyer, 475–488. New York: Cambridge University Press.

Cobb, M. D., and J. Macoubrie. 2004. Public perceptions about nanotechnology: Risks, benefits, and trust. *Journal of Nanoparticle Research* 6 (4): 395–405.

Dames, P., B. Gleich, A. Flemmer, K. Hajek, N. Seidl, F. Wiekhorst, et al. 2007. Targeted delivery of magnetic aerosol droplets to the lung. *Nature Nanotechnology* 2: 495–499.

Delgado, C., H. Short, and J. Krajcik. 2009. Design, implementation, and evaluation of the effectiveness of a 12-hour middle school instructional unit for size and scale. Paper presented at the National Association for Research in Science Teaching conference, Garden Grove, CA.

Fellows, N. J. 1996. A window into thinking: Using student writing to understand conceptual change in science learning. *Journal of Research in Science Teaching* 31 (9): 985–1001.

Glennan, S. 2002. Rethinking mechanistic explanation. *Philosophy of Science* 69 (Supplement to issue 3): S342–S353.

Krajcik, J. S., and C. Czerniak. 2007. *Teaching science in elementary and middle school: A project-based approach.* 3rd ed. New York: Lawrence Erlbaum Associates.

Krajcik, J. S., K. McNeill, and B. Reiser. 2008. Learning-goals-driven design model: Developing curriculum materials that align with national standards and incorporate project-based pedagogy. *Science Education* 92 (1): 1–32.

McNeill, K., D. J. Lizotte, J. S. Krajcik, and R. W. Marx. 2006. Supporting students' construction of scientific explanations by fading scaffolds in instructional materials. *Journal of the Learning Sciences* 15 (2): 153–191.

Mislevy, R. J., L. S. Steinberg, R. G. Almond, G. D. Haertel, and W. R. Penuel. 2003. *Leverage points for improving educational assessment* (PADI technical report no. 2). Menlo Park, CA: SRI International.

National Research Council (NRC). 1996. *National science education standards.* Washington, DC: National Academies Press.

Pellegrino, J., and members of the NCLT. 2008. Using construct-centered design to align curriculum, instruction, and assessment development in emerging science. Paper presented at the International Conference of the Learning Sciences, Utrecht, the Netherlands.

Rivard, L. P., and S. B. Straw. 2000. The effect of talk and writing on learning science: An exploratory study. *Science Education* 84 (5): 566–593.

Roco, M. C., and W. S. Bainbridge. 2006. *Nanotechnology: Societal implications II —Individual Perspectives.* Berlin: Springer.

Ruiz-Primo, M. A., M. Li, C. Ayala, and R. J. Shavelson. 2004. Evaluating students' science notebooks as an assessment tool. *International Journal of Science Education* 26 (12): 1477–1506.

Salmon, W. 1984. *Scientific explanation and the causal structure of the world.* Princeton: Princeton University Press.

Sham, J., Y. Zhang, W. Finlay, W. Roa, and R. Löbenberg. 2004. Formulation and characterization of spray-dried powders containing nanoparticles for aerosol delivery to the lung. *International Journal of Pharmaceutics* 269: 457–467.

Song, M., and C. Quintana. 2009. WIIS: Multimodal simulation for exploring the world beyond visual sense. Paper presented at the Conference on Human Factors in Computing Systems, Boston, MA.

Stevens, S., L. Sutherland, P. Schank, and J. S. Krajcik. 2009. *The big ideas in nanoscience.* Arlington, VA: NSTA Press.

Sung, J., B. L. Pulliam, and D. A. Edwards. 2007. Nanoparticles for drug delivery to the lungs. *TRENDS in Biotechnology* 25 (12): 563–570.

Ward, W. C., N. Frederiksen, and S. B. Carlson. 1980. Construct validity of free-response and machine-scorable forms of a test. *Journal of Educational Measurement* 17 (1): 11–29.

Wiggins, G. P., and J. McTighe. 1988. *Understanding by design.* Alexandria, VA: Association for Supervision and Curriculum Development.

Issues-Based Learning and Inquiry in Environmental Science:

Meeting the Third Goal of the National Science Education Standards

Jason Pilot and Doug Jones
Sir Winston Churchill Collegiate and Vocational Institute

Wayne Melville and Anthony Bartley
Lakehead University

Setting

As in many other education jurisdictions, environmental education in Ontario is infused throughout the science curriculum. This infusion requires students to not only understand the science behind environmental issues, but also be provided with "meaningful contexts for applying what has been learned about the environment, for thinking critically about issues related to the environment, and for considering personal action that can be taken to protect the environment" (Ontario Ministry of Education 2008, p. 38). In this chapter, we would like to share the experiences of one secondary school and its community in developing these "meaningful contexts."

Introduction

Sir Winston Churchill Collegiate and Vocational Institute, Thunder Bay, Ontario, is one of four public secondary schools operated by the Lakehead District School Board. From 2004 to 2008 the school offered environmental science as a locally developed grade 11 course. This course was superseded in 2009 by an Ontario Ministry of Education–developed course with the same name and focus. Despite this change, the focus of the course has consistently been the teaching of environmental science in the context of the issues that relate to the Northern Ontario region. These key issues are forestry, agriculture, waste management, energy conservation, and watershed management. Although traditional industries such as forestry are in decline, agriculture and waste management are beginning to establish themselves as viable industries. This accelerating change in the regional economy offers many opportunities for students to be involved in "public discourse and debate about matters of scientific and technological concern," the third goal of the National Science Education Standards (NSES), (NRC 1996, p. 13).

It is important to note that Sir Winston Churchill Collegiate and Vocational Institute is the only high school in the district with a long-term environmental science program. In recent years, the school has worked with community partners from Lakehead University (Roots to Harvest Program and Faculty of Science and Environmental Studies), government agencies (Ontario Ministry of Natural Resources and Lakehead Region Conservation Authority), local farmers (Mr. Arjen DeBruin and Mr. Kevin Belluz) and local merchants (Thunder Bay Grower's Market). These partnerships have been crucial to the initiation and success of the Environmental Science program.

The program draws upon students from all parts of the educational spectrum. The two courses offered to date have been at the workplace level (for students who will enter the world of work after high school), and the mixed level (students who will enter either college or university), classes which usually draw more applied learners than academic. The approach to delivering content in these courses has to strike a balance between applied and academic learning. To facilitate this balance, the teacher (Jason, the first author) chose to use issues-based learning and inquiry as the key foci for the teaching strategies and planning of the courses.

Jason has always used concrete issues in his classes as the catalyst, or platform, for learning content. The revised Ontario science curriculum, which was first implemented in 2009, recognizes the value of issues-based learning, a change that is highlighted in the Fundamental Concepts section of the grade 11/12 Ontario Curriculum: "Change the focus of the curriculum and instruction from teaching topics to 'using' topics to teach and assess deeper, conceptual understanding" (2008, p. 5). The revised curriculum also identifies "big ideas" as the broad points of understanding. These are expected to be retained by students after specific content has been forgotten. Jason uses these big ideas, or issues, to develop the context for students to understand the fundamental concepts. While they are identified in the curriculum, Jason has found that they have more impact on students when they are articulated by guest speakers or through media analysis.

The remainder of this chapter considers several key points that illustrate how an issues-based and inquiry-based course can promote "public discourse and debate about matters of scientific and technological concern." These include the development of a teacher's awareness of the local community context in which the class operates and a brief discussion of the philosophical foundation that serves to justify the use of the issue-based learning and inquiry strategies. For this chapter we define *inquiry* as grounded in the summation provided by the NSES:

> Inquiry is a multifaceted activity that involves making observations; posing questions; examining books and other sources of information to see what is already known; planning investigations; reviewing what is already known in light of experimental evidence; using tools to gather, analyze, and interpret data; proposing answers, explanations, and predictions; and communicating the results. Inquiry requires identification of assumptions, use of critical and logical thinking, and consideration of alternative explanations (NRC 1996, p. 23).

From this base, we explore the teaching and assessment and focus on how integration of the local community promotes considered student responses to local scientific issues. The chapter concludes with opinions of students and community partners as to the benefits of this approach.

Jason's Foundational Knowledge

An integral feature of the course has been the desire to simulate a real-world project that potentially impacts our local school community or a community in our region. From teachers who have tried to develop environmental science programs, Jason learned early on that there must be something special to attract students. From consulting with family and friends in various industries in Northern Ontario, he ascertained there were certain skills that students needed to be successful in any career: good communication (in various forms, especially verbal); knowledge of computer and information technologies; and most important, problem-solving skills were always at the forefront. Jason decided these skills needed to be the lure to bring students back to the environmental science program year after year. Only through using authentic, intense, and concrete issues as the backbone of student learning could this be achieved.

The Environmental Science Course in Outline

The environmental science course at Churchill has evolved since 2004. Originally, the course was based on locally developed curriculum materials obtained from school boards in Southern Ontario, tailored to meet the needs of Northern Ontario. In 2009, the Ontario Ministry of Education implemented a new environmental science course that encourages schools to use the strengths of their community to develop appropriate lessons and activities to cover the curriculum (Ontario Ministry of Education 2008). This new course was developed in response to the 2007 Provincial government report into environmental education, *Shaping Our Schools, Shaping Our Future*. This report was prepared by the Working Group on Environmental Education, which was headed by the Canadian astronaut Roberta Bondar and is more commonly referred to as the Bondar Report. We believe there is a clear resonance between the third goal of the NSES and the report's definition of environmental education:

> the responsibility of the entire education community. It is a content area and can be taught. It is an approach to critical thinking, citizenship, and personal responsibility, and can be modelled. It is a context that can enrich and enliven education in all subject areas, and offer students the opportunity to develop a deeper connection with themselves, their role in society, and their interdependence on one another and the earth's natural systems (p. 10).

As the environmental science course has evolved, Jason, in consultation with the other members of the science department, has sought to maintain a sense of coherence with other courses by the consistent use of two teaching approaches: issue-based learning and science as inquiry. Science as inquiry has been a consistent focus of the department since 2000 (see Jones et al. 2009). Coherence is achieved by planning each course toward a *culminating activity* that gives students opportunities to demonstrate their knowledge of the issue; the skills and abilities that they have developed through the course; and crucially, their capacity to engage in discourse about the issue. The culminating activity has varied over the years, including designing a greenhouse for our school community, planning a five-year forest management harvest plan for a regional provincial park, and designing a water or sewage treatment system for northern aboriginal communities.

The Environmental Science Course in Practice

Once the issue for the culminating activity has been determined, the learning expectations required by the curriculum are highlighted and used as the building blocks for the course. The units of work that develop from the expectations are not discretely sequenced one at a time, but rather they are woven into subunits based on different environmental issues (not necessarily the one used for the culminating activity) with overlapping expectations from multiple strands of the course. These learning expectations outline the core knowledge and skills of the course and are provided to the students at the start of the semester. This strategy helps to make assessment as transparent as possible, and also provides a framework for students to understand the direction and purpose of the course's components and how they are woven together throughout the semester. We believe it is important that as students progress, they are aware that they are using resources and learning techniques they will be called upon to apply at the end of the semester. A key example of this is information gathered by visiting DeBruin's Greenhouses. Students acquire knowledge of how to maintain a greenhouse to the specific characteristics of the crops being grown in it. They complete information sheets as the farmer gives the class a breakdown of his operation for each crop (tomatoes, cucumbers, wheat grass, and peppers). These sheets are used by students during their culminating activity. Another example is creating a "cradle-to-grave" study of a consumer product to determine the ecological and carbon footprint. The complex operations of producing a consumer product can be directly translated to reducing a community's ecological and carbon footprint by developing a community garden or tertiary treatment facility for a water treatment plant. This approach to teaching is analogous to that taken by coaches of any sport. Students need to acquire knowledge and skills throughout the semester, constantly practicing and honing those skills so they can apply them to the culminating activity, or as Jason refers to it, "the big game day."

The course is delivered in a manner that balances subject knowledge and its application. For example, students collect tree stand data in the forestry inquiry or alter the abiotic conditions of soil in an agricultural inquiry. When students study energy consumption and alternatives, they build wind turbines and solar cookers and determine the power consumption of each device as they compete to build the best prototypes. These activities are integrated throughout the course and their importance cannot be overstated. Students need to get their hands dirty. They need to feel connected to the topics they are learning about and feel a sense of accomplishment. Technological problem-solving activities and scientific inquiries develop the thought processes required for the culminating activity and give Jason a chance to see which people work well together. It is interesting to observe the group dynamics when the class is given a difficult problem to solve and they are competing with their peers for the best prototype. Similarly, open inquiries allow students to demonstrate their practical lab skills, researching abilities, or communication skills.

Due to the large emphasis the science department puts into inquiry from grade 9 to 12 at Churchill, Jason's students have excellent inquiry skills by the time they arrive in grade 11. This is an important point to ponder in light of one of the objections that is often raised to the teaching of science as inquiry: that it is too time-consuming (Anderson 2002). One result is that inquiry is not common in schools, with the majority of high school graduates leaving school "without even one experience with real science" (Yager 2005, p. 20). Our experience indicates that developing science as inquiry across the department, from the early days of secondary school, pays a dividend in saving time and effort in later years.

From the beginning of the course, several inquiries—ranging from guided to completely open expose students to situations similar to what they will encounter in their culminating activity. Examples of inquiry topics include, "What factors affect the production of wheat grass?" and "What factors affect the remediation of contaminated water?" The topics for the open inquiry vary depending on the time of year, available resources, and chronology of units covered in the semester. Within each topic, each group conducts an open inquiry in which they select their own research question (i.e., "How do soil nitrogen levels affect the production of wheat grass for juicing purposes?"), develop their own methodology, complete their testing and then publicize their inquiry in both a written document and a conference-style presentation. Each group uses video footage to show how the inquiry was carried out, what their results were, and what their analysis leads them to theorize about the relationship among their independent and dependent variables. The other student groups, acting as the "scientific community," challenge the presenters on their methodology and analysis. The teacher evaluates the presentation, questions posed to other groups, how well the presenting groups can defend their work, how well they understand the inquiry process, and how well they conduct an inquiry in the lab (see Appendixes A and B, pp. 228–230). A major assessment role for the teacher is to balance evaluation both academic and applied knowledge and skills for these large inquiries.

Meeting and Assessing the Third Objective

The skills and knowledge required to produce potential solutions to science-based issues and to communicate those solutions to a critical audience are both crucial foci building toward the culminating activity. Since 2004, however, Jason has found that skills and knowledge of themselves are insufficient to motivate students. A major source of student motivation to achieve the expectations is direct exposure to an issue in order to solve an authentic problem with real-world consequences. In meeting the expectations, they also "engage intelligently in public discourse and debate about matters of scientific and technological concern." Over the past three years the extent of that public discourse has expanded significantly. Each year the Culminating Activity has been established to further meet the needs of the students, the curriculum and a contemporary issue in the local community.

The provincial course takes students through five units of study: Scientific Solutions to Contemporary Environmental Challenges, Conservation of Energy, Human Health and the Environment, Reducing and Managing Waste, and Sustainable Agriculture and Forestry. Through each unit of study, key concepts that are related to either culminating activities (students choose either an agricultural or waste management) are highlighted for future reference. In a synopsis of one of the culminating activities, developing an agricultural plan for a northern aboriginal community, we can identify the many concepts and skills required to complete the task: Groups of three students use the skills and knowledge they've developed throughout the course to complete the activity. They draw upon their experiences growing wheat grass or tomatoes, trips to the local grower's market, information gathered from guest speakers such as local farmers, inquiries carried out during key units of study, and research skills honed from inquiries and research activities since grade 9. Exemplars are available from Churchill's science website: *http://swc.lakeheadschools.ca/science*.

Jason has worked with science chair Doug Jones (the second author) to develop his assessment strategies. Doug developed the original strategies for his grade 9 inquiry culminating activity. Jason has divided the assessment into three parts; expert group component, group report, and interview. Because this activity hinges on the effort of all participants, it is important to weigh both the understanding of individuals with the quality of work and depth of coverage of each section (see Figure 14.1).

Figure 14.1. Expert Group Tasks

Snapshot of the Community	Methods of Agriculture and Nutrient Management
Give an overview of what is available to the people of the community in terms of fresh produce and vegetables. Include the cost of eating these foods. What is the topography surrounding the community? Locate suitable areas for farms and greenhouses. Identify how goods are imported and exported from the community. What is the population of the community, its local economy, and its culture?	Types of greenhouses that can be utilized for a community like yours. Include the costs and approximate the cost of transportation. Different methods of outdoor farming and pertinent information on how to farm properly (like crop rotations, pesticide use, fertilizing, organic gardening, nutrient management *etc.*). Include the cost of each method. Soil types of the area and any deficiencies in nutrients or texture that need to be addressed. Methods of reducing energy and raw material consumption in that geographic area.
What Can We Produce?	**Explanation of Marking Scheme**

Research important information on the climate patterns and important factors to growing seasons;
  -range of monthly temperatures (highs and lows)
  -amount of sunlight/month
  -length of outdoor growing season
Species of vegetables and grains that can be grown in a greenhouse setting and outdoor farm. Which ones are annuals and perennials?
What are the nutritional benefits of each species?
Determine the different uses for each type of produce.
Determine where seeds would come from and the cost of acquiring the seeds.
	Expert Group Component: 30 %
	Group Report: 40 %
	Interview: 30 %
	Total: 100%

This activity requires group members to complete specific tasks; however, the group must synthesize the parts into a cohesive whole. The expert group component, which is worth 30%, is evaluated using a scoring guide (rubric) to rate the depth of coverage of each specific component presented in the report, what sources of research were used and the understanding of each component. The next two parts, report and interview, are evaluated using rubrics to assess the quality of work or understanding of the overall report (see Appendixes A and B, pp. 228–230). The report rubric is based on the achievement charts from the 2009 Science 11/12 Curriculum Documents developed by the Ontario Ministry of Education (2008). See *www.edu.gov.on.ca/eng/curriculum/secondary/science.html*.

The categories that Jason has chosen for assessment are communication and application. This rubric assesses how well the report was designed and also communicates student ideas and how well the connections have been made between their knowledge and their research. The interview rubric assesses degree of prompting, verbal communication, and knowledge (see Appendix C, p. 231). The interview itself is 15 minutes long and groups submit their reports online to the teacher 24 hours prior to the interview. They are given certain prescribed questions when they begin the culminating activity. Students are responsible for understanding each other's work and how it fits into the final report. In addition there are impromptu questions that arise based on individual student answers or from the group reports. There are six prescribed questions and two or three impromptu ones. All group members can take part in answering the impromptu questions and individual credit is given. (For an example of a report see Appendix D, p. 232.)

A final test is given during the last week of classes to evaluate how much of the core content or knowledge each student possesses. Due to the variety of students in the classes, most students do much better in the culminating activity than the written test.

What Others Think: Students and Community Partners

The environmental science program at Churchill has had great success. Since 2004, the issues-based and inquiry foci of the course have engaged students with issues that are of concern to them and promoted the pursuit of science-based careers at both the university and college levels. Conversations with former students provide anecdotal evidence that the program prepared them well for the first years of their postsecondary education. For example, many of the identification and field data collection activities are similar, if not identical, to those they completed in high school. Also, students' higher-level problem-solving skills are those laid out in the revised curriculum documents as being paramount to developing life-long learners in both postsecondary education and the world of work.

For this chapter, Jason interviewed both current and former students. The consensus appears to be that students find these activities very rewarding because they are creating documents that could actually be used to improve the lives of others. Student testimony like the one below is commonplace:

> I found the field trips great. The more work we could do outside the classroom the better. I thought the projects were fun. We had an input into something (the agricultural culminating activity), and the work inside the classroom could be used to help the community. I really liked the fact that I was doing a project on my community (Thunder Bay) that could be researched within the community. From this I got interested in what I could do for my community. This class has sparked my interest in environmental studies. I've learned that environmentalism isn't just about fundraisers but action, science and making your community better. I think to make the class better we need to get involved in projects, like community gardens, that get the students working after the class is over.

Jason also had the opportunity to discuss how their semester had prepared them for their studies. One student, who is attending Confederation College's Environmental Technician program, submitted the following response:

> I found that all of the hands on activities like the river studies and the mapping we did were the same as the activities I did in my field classes. I had no problem falling into the activities. What we did in class was bang on what I did in first and second year classes.

During this student's years the focus was on watershed management and forestry. The same focus on issue-based learning and development of practical skills was used during his first two years of postsecondary study.

During the last two years our program has partnered with an agency from Lakehead University called Roots to Harvest. This program was established to develop and promote awareness of Food Security at all educational levels and within the community. Our students took part in practical activities like growing crops at a local farmer's greenhouses to sell in the school and local market, and agricultural inquiries at Lakehead University's greenhouses. In discussions with our community partners Jason has had a chance to get their perspective on the benefits of this program:

> From start to finish, this inquiry project brought out the best in these students, of all different academic levels and experiences. Using an inquiry approach, Jason created a situation where students could be successful and produce high level end products that reflected all of the learning they had gained through previous knowledge, their own life experiences, research, problem solving and innovation. The students truly internalized and processed the aspects of food security that most people only glimpse at the shallowest level.
>
> I have learned so much from these opportunities to work with Jason and his classes. With 15 years of my own experience of being an experiential educator, I have never seen students of all abilities gain so much from an academic experience as I have with these projects in Jason's classes.
>
> — *Erin Beagle (head of Roots to Harvest)*

and;

> I would love to see even 1% of students that visit our farm in Jason's classes working for a grower or start their own (agricultural) business. The community of Thunder Bay is under-developed for growing fresh produce. Our greenhouse growing needs to be expanded. The "hands-on" learning is the most important part of learning in my mind. It is very important that Jason continues to bring classes to farms in the community. Learning on a farm and learning by doing hands on activities is not like learning in a classroom. There's no chance for Jason's kids to fall asleep on my farm.
>
> —*Arjen DeBruin (Owner of DeBruin's Greenhouse's)*

There is no question that students look forward to the practical activities in this program. It continues to be a recruiting tool for a program that started with only 12 students in one class and within five years has grown to four classes being offered each year.

Evidence of Impact

The program has a major impact on the school community and, increasingly, on the wider community. The school has been one of the few in the region to complete annual waste audits. Students in environmental science classes, and recently in student groups, have taken waste reduction to the forefront of student and faculty consciousness. In 2008, as part of the agricultural component of the course, students grew and produced wheat grass for sale within the school and at the local grower's market. The class developed a business selling wheat grass shakes and shots. This business generated enough money to purchase a juicer, pay for equipment, and support the purchase of a greenhouse for the school. During the year, the class had members of the farming community come to the classroom and also visited farms, commercial greenhouses, and the Lakehead University biology department to learn about local farming and markets. These activities were reported by the Canadian Broadcasting Corporation.

In 2009, the class took part in a 100K Day where they prepared food from ingredients that was produced within 100 km of the school. This activity was a response to the increasingly important issue of food security reported on in the local newspaper and has now spread to other schools across the city. The school will begin to hold 100K Days in the fall and spring each year, using food grown at the school and within the community. It is hoped that this program will expand to involve students in business, family studies, aboriginal studies, and other science classes. The continued development of the school's community garden and greenhouse will allow the cafeteria to serve produce grown at the school. The need for fresh produce that is affordable has been identified by the staff, students, and cafeteria personnel for many years now. It is hoped that with the environmental science program leading the way, Sir Winston Churchill Collegiate and Vocational Institute will continue to establish itself as a leader in healthy and environmentally aware living.

Conclusion

Since 2004, Grade 11 environmental science at Sir Winston Churchill Collegiate and Vocational Institute has promoted both issues-based and science as inquiry. The use of these strategies has helped many of our students to consider and pursue science-based careers. In a region that is undergoing fundamental changes in its economic structure, this is an important consideration. More importantly, the course has given students opportunities to consider and act on issues that are of interest to them. For students to be able to discuss issues with business and industry groups, educational institutions, and the media indicates that they are meeting the third goal of NSES: the capacity to "engage intelligently in public discourse and debate about matters of scientific and technological concern."

References

Anderson, R. D. 2002. Reforming science teaching: What research says about inquiry. *Journal of Science Teacher Education* 13 (1): 1–12.

Jones, D., C. Kaplanis, W. Melville, and A. Bartley. 2009. Science as inquiry at Sir Winston Churchill Collegiate and Vocational Institute. In *Inquiry: The key to exemplary science*, ed. R. E. Yager, 151–176. Arlington, VA: NSTA Press.

National Research Council (NRC). 1996. *National science education standards*. Washington, DC: National Academies Press.

Ontario Ministry of Education. *Ontario Curriculum, Grades 11 and 12: Science, 2008 (revised)*. 2008. Toronto, ON. Queens Printer for Ontario.

Working Group on Environmental Education. 2007. *Shaping our schools: Shaping our future*. Toronto, ON: Queen's Printer for Ontario.

Yager, R. E. 2005. Achieving the staff development model advocated in the national standards. *Science Educator* 14 (1): 16–24.

Websites

The Ontario Ministry of Education: *www.edu.gov.on.ca/eng/curriculum/secondary/science.html*
Sir Winston Churchill Collegiate and Vocational Institute: *http://swc.lakeheadschools.ca/science*

Appendix A

Grade 11 U/C Environmental Science: Water Filtration Inquiry Evaluation Sheet

Group Members: _____

Planning/Executing the Inquiry [20%]

Level 1	Level 2	Level 3	Level 4
☐ Formulates questions and develops hypotheses with limited effectiveness	☐ Formulates questions and develops hypotheses with some effectiveness	☐ Formulates questions and develops hypotheses with considerable effectiveness	☐ Formulates questions and develops hypotheses with a high degree of effectiveness
☐ Performs inquiry including observing, manipulating materials, gathering data and using equipment safely with limited effectiveness	☐ Performs inquiry including observing, manipulating materials, gathering data and using equipment safely with some effectiveness	☐ Performs inquiry including observing, manipulating materials, gathering data and using equipment safely with considerable effectiveness	☐ Performs inquiry including observing, manipulating materials, gathering data and using equipment safely with a high degree of effectiveness

Communication/Critical Thinking Skills (Report) [40%]

Level 1	Level 2	Level 3	Level 4
☐ Uses conventions, vocabulary and terminology in the formal lab report discipline with limited effectiveness	☐ Uses conventions, vocabulary, and terminology in the formal lab report discipline with some effectiveness	☐ Uses conventions, vocabulary, and terminology in the formal lab report discipline with considerable effectiveness	☐ Uses conventions, vocabulary, and terminology in the formal lab report discipline with a high degree of effectiveness
☐ Analyzes and interprets data with limited effectiveness	☐ Analyzes and interprets data with a some effectiveness	☐ Analyzes and interprets data with considerable effectiveness	☐ Analyzes and interprets data with a high degree of effectiveness
☐ Evaluates analysis and research to formulate and justify conclusions with limited effectiveness	☐ Evaluates analysis and research to formulate and justify conclusions with some effectiveness	☐ Evaluates analysis and research to formulate and justify conclusions with considerable effectiveness	☐ Evaluates analysis and research to formulate and justify conclusions with a high degree of effectiveness

(Appendix A *continued*)

Critical Thinking/Processing Skills (Presentation and Questions) [40%]

Level 1	Level 2	Level 3	Level 4
❑ Communicates relevant information pertaining to steps taken in scientific inquiry with limited effectiveness	❑ Communicates relevant information pertaining to steps taken in scientific inquiry with some effectiveness	❑ Communicates relevant information pertaining to steps taken in scientific inquiry with considerable effectiveness	❑ Communicates relevant information pertaining to steps taken in scientific inquiry thoroughly and considerable effectively
❑ Demonstrates a limited understanding of scientific validity with respect to errors and accepted observation skills	❑ Demonstrates an adequate understanding of scientific validity with respect to errors and accepted observation skills	❑ Demonstrates a good understanding of scientific validity with respect to errors and accepted observation skills	❑ Demonstrates an excellent understanding of scientific validity with respect to errors and accepted observation skills

Appendix B

Grade 11 Environmental Science Culminating Activity-Evaluation Sheet Agricultural Plan

Individual Evaluation _____ / 30

Name: _____ Name: _____ Name: _____

Snapshot of Community	Proposed Plan for Community	What Can We Produce?
Give an overview of what is available to the people of the community in terms of fresh produce and vegetables. Include the cost of eating these foods. 0 1 2 3 4 5 6 7 8 9	Types of greenhouses that can be utilized for a community like yours. Include the costs, and approximate the cost of transportation. 0 1 2 3 4 5 6 7 8	Research important information on the climate patterns and important factors to growing seasons; - range of monthly temperatures (highs and lows) - amount of sunlight per month - length of outdoor growing season 0 1 2 3 4 5 6 7 8 9 10
What is the topography surrounding the community. 0 1 2 3 4	Different methods of outdoor farming and pertinent information on how to farm properly (like crop rotations, pesticide use, fertilizing, organic gardening, and nutrient management etc.). Including the cost of each method. 0 1 2 3 4 5 6 7 8 9 10	0 1 2 3 4 5 6
Locate suitable areas for farms and greenhouses. 0 1 2 3 4	0 1 2 3 4 5 6	Species of vegetables and grains that can be grown in a greenhouse setting and outdoor farm. Which ones are annuals and perennials? 0 1 2 3 4 5
Identify how goods are imported and exported from the community. 0 1 2 3 4 5 6 7 8 9	Soil types of the area and any deficiencies in nutrients or texture that need to be addressed. 0 1 2 3 4 5 6	What are the nutritional benefits of each species? 0 1 2 3 4
What is the population of the community, its local economy, and its culture?	Methods of reducing energy and raw material consumption in that geographic area.	Determine the different uses for each type of produce. 0 1 2 3 4 5
		Determine where seeds would come from and the cost of acquiring the seeds.

Group Evaluation

Criteria	Level 1	Level 2	Level 3	Level 4
Knowledge • understanding of relationships between concepts	☐ demonstrates limited understanding of relationships between concepts	☐ demonstrates some understanding of relationships between concepts	☐ demonstrates good understanding of relationships between concepts	☐ demonstrates excellent understanding of relationships between concepts
Making Connections • analysis of social and economic issues • proposing courses of action based on scientific principles	☐ analyzes social and economic issues with limited effectiveness ☐ extends analyses of familiar problems into courses of practical action with limited effectiveness	☐ analyzes social and economic issues with some effectiveness ☐ extends analyses of familiar problems into courses of practical action with some effectiveness	☐ analyzes social and economic issues with good effectiveness ☐ extends analyses of familiar problems into courses of practical action with good effectiveness	☐ analyzes social and economic issues with excellent effectiveness ☐ extends analyses of familiar problems into courses of practical action with excellent effectiveness
Communication • communication of ideas and concepts	☐ communicates ideas and information with limited clarity and precision	☐ communicates ideas and information with some clarity and precision	☐ communicates ideas and information with good clarity and precision	☐ communicates ideas and information with a high degree of clarity and precision

Appendix C

Interview Evaluation

Criteria/Perform	1	2	3	4
Degree of Prompting Required	Much prompting or rewording of the question is needed. The student spends much of the time in the group report.	Some prompting or rewording of the question is needed. The student sometimes relies on the group report.	Virtually no prompting or rewording is necessary. The student does not refer to the group report.	No help is required. Reference is made to the report but it is not opened.
Communication Skills: Clarity, Diction, Gestures, Eye Contact	The response is broken. Ums, Ahs, etc. No or excessive gestures. Little eye contact	The response has some breaks. Some diction and grammar errors are made. Gestures may be excessive. Eye contact may be limited.	The response, diction, grammar, gestures and eye contact are satisfactory.	The response is smooth flowing, diction and grammar are good. Eye contact and gestures aid the answer.
The Degree To Which The Question Is Answered	An attempt is made to answer some of the question.	Much of the question is answered.	All parts of the question are addressed in a satisfactory manner.	The question is completely and eloquently addressed.

Student Name: _____

Question #	Degree of Prompting	Communication	Degree of answer
1			
2			
3			

Student Name: _____

Question #	Degree of Prompting	Communication	Degree of answer
1			
2			
3			

Student Name: _____

Question #	Degree of Prompting	Communication	Degree of answer
1			
2			
3			

Overall Mark

Student	Checklist (20)	Report (50)	Interview (30)	Final Mark

Appendix D—Example of Report Agricultural Plan

Snapshot of the Community

The topography surrounding Thunder Bay in general is the result of continental uplift and erosion, the movement of glaciers, and the continuing actions of wind, waves, water, and gravity. In the north, there are the Hudson Bay lowlands, which are mostly· swamp, meadow and forest. The <u>Canadian Shield</u> covers the rest of northern Ontario, and extends into the southeast. The Great Lakes-St. Lawrence River Lowlands in southern Ontario provide fertile soils and ideal farmland.

Several suitable areas were located for growing food; we thought putting a greenhouse in would be best, as the soil in the area isn't extremely wonderful. A possible location for the greenhouse would be at the Ogden Community Centre (600 McKenzie Street), which this way, would be more well-known to the general public. Depending on the condition of the river, wild rice could be planted in the Neebing-McIntyre floodway. This has been done in previous years. McKellar Island is another potential site for establishing a farm. There is a vacant lot there, but of course it depends how fertile the soil is. All of these locations are probable, but would need further looking-into.

As far as imports and exports go, this part of Thunder Bay does what the rest of the city does. The majority of imported products come in by truck, or sometimes by train. There aren't many large retail, grocery, or other miscellaneous stores in this area.

The population of the East End, as of 2007, is about 5,000. This was discovered by taking a census. The Business Improvement Association (BIA) recently planted trees, hung plants, and installed benches to improve the appeal of the area. The area is mostly made up of residential properties; most are small single homes… most of these were built before 1946. The East End also has a public pool, a community centre, and two elementary schools. Other than the school yards and play areas adjacent to the pool and community centre, the neighbourhood has no other parks or public green spaces. Low income levels increase from north to south in the neighbourhood, from 16.3 to 22.8 to 35%, compared with a regional rate of 11.4%. The neighbourhood is mainly populated by the descendents of European immigrants (Polish, Ukrainian, British, German, Finnish, Dutch, and Italian) and by a high proportion of aboriginals.

Methods of Agriculture

When building a greenhouse for the East End community, the greenhouse needs to fit the basic needs of the area. The type of greenhouse that can best suit this is the Warm Greenhouse. This greenhouse can grow almost any type of regular garden vegetable or flower plant and can still be used in the winter months if heat and growing lights are used. Using a Hothouse Greenhouse or a Cool Greenhouse will not suit the needs of the East End community and will not be able to grow the types of plants needed. The Hothouse Greenhouse is used for growing tropic and exotic plants, and the temperature is kept at 65°. This can become very expensive if the greenhouse is used year-round, whereas the Warm Greenhouse is much cheaper and only needs heat and growing lights during the winter months. The Cool Greenhouse is cheaper but is not an option for the East End community because of the types of plants it can grow. A Cool Greenhouse can only grow seedlings and plants that do not need high heat to survive, which is not the type of plants needed for the East End community. Creating a community garden is also not a possibility because the soils in the East End community are not suitable for growing plants. The soils have bad drainage and may flood the plants. This area is only suitable for growing swamp or marsh plants and does not have enough loam soil to grow garden type plants for this community.

To properly farm and continuously grow plants throughout a year, crop rotations for growing an organic garden are extremely important. An organic garden is a garden that does not use pesticides, unnatural fertilizers, or added nutrients and does not genetically modify its produce; or a garden that grows natural organisms. For an organic garden, crop rotations can be used in place of pesticides that kill annual pests or host specific pests, such as hornworm or fusarium that only prey upon solanaceous crops (such as tomatoes, potatoes, eggplant and pepper). Crop rotation means changing the place of your plants and growing different plants every year. This will cause the disease or pest to die off or relocate to a different farm because its food source is gone. Labeling which plants you had in which section of your farm every year is a good way to remember not to plant something in the same area. Fertilization is also very important for organic gardening, but the source of the fertilizer is very important if you want your farm to be exclusively organic. Farm-produced fertilizers are the best source unless the manure producer was fed inorganic feed. Cows that ate genetically modified produce do not produce organic manure and cannot be put in a garden if the farmer wants it to be organic. Plenty of research is needed when using fertilizer for an organic garden. Depending on the types of plants the garden is growing, nutrient management is an important thing to consider. Certain plants need certain types of nutrients and fertilizers to grow properly and can only grow with the right amount of these nutrients.

Soil and Nutrient Management

The soil of East End is classified as a combination of Mission and Fort William soil. Mission soil is a mixture of sand, sandy loam, and peaty phase. It is stone free but the drainage is poor. Fort William soil is a mixture of loam, sand, and sandy loam. It is also stone free and drainage is imperfect. Altogether the soil is relatively moist because of the lowness of the ground to water level. East End is a flat area with poor drainage despite its sandy soil because of the soils' closeness to the water table underground, preventing the water from going anywhere else. These conditions are imbalanced for the growth of agriculture within the community.

Soil types of the area are very important if planning on having a farm or greenhouse/garden to support the community. For growing vegetable plants, loamy soils are favored as the best for growing plants. Loam contains equal parts sand, clay, and silt. It is easily tilled and retains moisture and is well-balanced. Loam soil is amongst the most fertile soils and almost any crop can be grown in it. Sandy soil is easy to cultivate and warms up quickly in spring. It drains well so the plants do not stand with their roots in water for too long. However, as it drains quickly, plants need to be regularly watered and fed if they are to thrive. Adding organic material to supplement sandy soil is best for gardening. Otherwise, the water can run through it so quickly that plants won't be able to absorb it and the nutrients.

The combination of sand and loam in the East End creates relatively good growing soil but is slightly acidic. The soil contains fewer nutrients than other soils and is prone to over-retaining water. Through good management and use of fertilizer and artificial drainage, excellent plants can be grown. With the use of topsoil and organic matter as nutrients, the soil can be pruned for specific vegetables to feed the community. Adding organic matter to a sandy soil increases its water-holding capacity and improves its fertility. For farms, soils without good drainage or with too much drainage can prevent plants from growing properly or getting enough nutrients. Not enough drainage can cause plants to flood and too much drainage can kill the plants with dehydration.

When planning to plant in these types of soils, the ground should first be loosened and turned over and then a 1- to 2-in protective layer of mulch on the soil surface above the root area. Cultivating and mulching reduce evaporation and soil erosion. To help prevent too much draining, a raised bed can be constructed using railroad ties, construction timbers or other suitable materials and then additional topsoil can be added with necessary nutrients like nitrogen, phosphorus, potassium, and sulphur. This soil is partially potassium-deficient and needs extra nutrients from sources like manure and compost. To lower the amount of acidity of the soil, lime can be added but is not always recommended for use. To ensure best results, the nutrients should be placed directly on the soil and not on the plants to prevent harm to the leaves.

An "HOLA" Approach to Learning Science

Theodora Pinou
Western Connecticut State University

Marjorie Drucker
Barnard Environmental Studies Magnet School

Elizabeth Studley
Solar Youth

Setting

Barnard Environmental Studies Magnet School, a grades preK–8 school, is located in an urban New Haven district in Connecticut. It is bordered by the West River, which empties into Long Island Sound. Environmental studies is the connecting theme that weaves throughout the curriculum. Overarching environmental units connect the core curriculum to schoolwide studies of local rivers and organisms that live in them. Barnard Environmental Magnet School has partnerships with local specialists that work with teachers and students to encourage outdoor educational experiences that enrich these environmental themes. The school's website is *http://schools.nhps.net/barnard*.

Barnard is part of the New Haven Public School Interdistrict Magnet School Program. Our magnet program draws from 28 different towns outside of New Haven, Connecticut, with the goal of providing racial, ethnic, and economic diversity throughout the magnet schools. Barnard School families apply through a lottery system that is open to the public. Each of the district's magnet schools offers a unique curriculum and many specialized programs. Barnard school's magnet theme is environmental studies. We use the Connecticut State and district science standards as the basis for the curriculum developed at the school, and these standards are aligned with the national science standards. Four overarching themes (freshwater, energy, migration, Long Island Sound) connect the core curriculum to schoolwide studies of our local environment. Thus, fifth-grade students focus on: Energy and Climate Change; Light and the Environment; and Water Conservation and Pollution.

Introduction

The physical design of the school includes outdoor gardens, greenhouses, and a nature center that is adjacent to the West River Memorial Park. Students access the nature center through a pedestrian bridge that spans a busy boulevard. Each grade is associated with unique experiences that contribute to the development of environmental stewardship. An example of this is the return of Atlantic salmon fry to Eight Mile River by grade 3 students, population studies of Long Island Sound creatures aboard the R.V. Island Rover by grade 4 students, and birding by canoeing in The West River by grade 6 students. Each grade is associated with a specific animal. The animals also act as a theme through which unique, grade-specific environmental studies develop. For example, grade 1 studies salamander development in relation to temperature change; grade 2 grows butterflies from caterpillars; and grade 4 studies the migratory patterns of sea turtles. All of these experiences bring to the student a deeper understanding of the interconnection between humans and all living things. They understand the impact that human actions can have on the delicate balance between humans and organisms with which they share the planet. The combination of this school program and exposure to practicing biologists and ecologists permits students to continuously engage in conversation with experts and collaboratively consider ways they can care for and sustain their community resources (such as gardens, parks, rivers, and wetlands).

The many studies students conduct at all grade levels act to heighten student awareness of local urban, as well as greater Connecticut issues. Students who attend Barnard for their entire K–8 experience will have considered the relationship between clean water and fish raised in chilled tanks. They will connect the release of these fish into the Connecticut River and healthy eating. They will have considered how a changing climate will impact Long Island Sound diversity as they ride aboard the RV Island Rover. They will realize that they can grow their own food and vegetables through the schools "Grow Box" and "Courtyard Garden" Programs and be empowered to sustain a healthy diet for themselves and their community. They will have experienced using alternative renewable energy such as solar panels found on their school and CFL lightbulbs utilized at their Smart Living Center. Thus, Barnard students gain practical experiences they can talk about when they need to consider solutions to current and relevant global issues. These practical experiences permit them to frame questions and responses as they engage in conversation and critical thought to support their varying ideas.

Differentiating Learning

Another advantage to collaborating with school partners is the ability to differentiate instruction for a diverse student population. Barnard partners bring diverse teaching practices into the classroom that offer students a variety of cognitive learning experiences, and permit classroom teachers to observe, assess, and reflect on the learning of their students. Therefore, although school performance in Connecticut is unidimensionally assessed (i.e., Connecticut Mastery Tests, CMT), infusing in-class partnerships into the curriculum can permit domain-specific giftedness to flourish, especially physical and intuitive/intrapersonal giftedness often hidden among high-context learners, a student population that typically represents urban demographics today (Gardner 1999; Sanchez-Burks et al. 2000; Ibarra 2001).

A student's ability to engage in thematic discourse and conversation with peers and practitioners demonstrates his/her ability to synthesize information and develop student-centered questions and responses that are relevant to student interests. This strategy permits the consideration of open-ended questions that students construct themselves and respond to. Such open-ended problems typically have more than one correct answer. A teacher can use student discourse and peer-to-peer conversation to develop instruments for alternative assessments that can be keys for accurately evaluating cognitive learning of a diverse student population. The urban context of the Barnard Environmental Magnet School makes alternative assessments valuable because they promote equity, increased motivation, and higher standards. They also empower students and teachers, and focus attention on teaching and learning (Winking and Bond 1995).

Supporting the National Science Education Standards

United States education reform policy for science education requires the implementation of national standards and goals (NRC 1996; 2000). The goals for school science, framed by the National Science Education Standards (NSES), define a scientifically literate society and are sensitive to the learning needs of all learners. While science literacy focuses on the mastery of content knowledge, it also focuses on using that knowledge to achieve social good (King 2007).

Goal 2 of NSES focuses on helping students use appropriate concepts and processes in making personal decisions. Goal 3 specifies that students should "engage intelligently in public discourse and debate about matters of scientific and technological concern" (NRC 1996, p. 13). The Barnard Environmental Magnet School's focus on environmental learning themes that lend themselves to conservation, human health, and resource management practices permit student engagement that is practiced through student-centered inquiry experiences. These are often guided by partnering with content specialists, permitting classroom teachers to use alternative assessments to measure student learning through questioning, community stewardship, and peer-to-peer teaching.

Implementing classroom instructional strategies that lend themselves to both unidimensional assessments and alternative assessments provide evidence of proficiency that can satisfy both school and student learning evaluations. Barnard Environmental Magnet School's collaboration with Solar Youth offers one such learning and assessment opportunity. Solar Youth's HOLA method of instruction uses hands-on inquiry strategies in combination with kinesthetic-based cognitive questioning and is the basis for this chapter.

Collaborating With Solar Youth's In-School HOLA Program

Solar Youth, Inc. is a nonprofit environmental education and youth development organization (*www.solaryouth.org*). Its mission is to provide opportunities for young people to develop positive senses of self and connections and commitments to others through programs that incorporate environmental exploration, leadership, and community service. The Hands-on Outdoor Learning Adventure (HOLA) Program is Solar Youth's in-school program that works in collaboration with New Haven Public Schools to support classroom achievement of the Connecticut State Core Science Curriculum Framework by supplementing teachers' curriculum with hands-on, outdoor lessons. Each participating class takes part in a two-hour, hands-on lesson once every two months, from October through May.

HOLA integrates music, kinesthetic learning, and naturalistic learning through a variety of teaching styles incorporated in the lesson. Using songs, creative movement, games, and hands-on activities, each lesson promotes creativity and enhances environmental stewardship. Practice in being a steward of the environment at an early age, along with adventure experiences and outdoor explorations, enhances the likelihood that stewardship will become a part of a child's daily lifestyle and persist throughout adolescence and adulthood. Dr. Greg Place states, "Time spent outdoors during an individual's early-life, with family, through organizations and education, whether positive or negative, helps impact adult's views toward the environment" (2004).

HOLA is not modeled after any program. It is an original strategy developed by Solar Youth's Director's commitment to help urban children learn through doing.

Learning Through Action and Conversation

Solar Youth facilitates several hands-on activities in their in-class program to teach content prior to formal assessment of included information in the curriculum and other instructional materials. These activities always involve conversations and discourse between and among peers, as well as between students and the school community (teachers, parents, and administrators). Such activities include outdoor field experiences (such as collecting leaves and then identifying them); games like Conservation Jeopardy (Figure 15.1), in which the students write the questions and then pose questions to their peers; and community poster presentations (Figure 15.2) where students report information to their peers. Activities that promote stewardship are also important for developing engagement with learning. Examples include park cleanup with park rangers and school loudspeaker announcements that highlight upcoming events and news.

Figure 15.1. Playing Conservation Jeopardy

The HOLA approach gathers students in an informal whole-group setting and uses repetition of terminology and sustained kinesthetic activities to measure learning and retention and use of central concepts and themes. The process implements "P(reviews)" as a way to measure students' knowledge acquisition and retention. With P(reviews) students answer the same set of questions at the beginning of a session (pre), at the end of the session (post), and at the beginning of the next session (post2). For example, if there are six questions at the beginning of a session, three are from the previous session and three are new. Students are assured at the beginning of a session that it is acceptable not to know the answers of the new questions. When asking review questions from a previous session, instructors always ask students if they remember what they learned together previously. Instructors jog student memories but do

Figure 15.2. Students Presenting an Activity to Their Peers

not go into a full-scale review. The kinesthetic singing and dancing that occurs simultaneously to the questioning helps to minimize the tendency to copy each others' answers and reminds students of the ethical importance in thinking for themselves.

Instructions for Conducting a "P(review)"

- Gather the group in a circle
- Hand each student a set of ABC letters
- Sing the Preview/Review song:

"Preview, Review
Preview, Review
Turn around....
Drop your letters to the ground....
Now freeze!
And listen to the question.....

(Ask the question.)

Is it A, or B, or C?
Tell me
Is it A, or B, or C?
Tell me
Now freeze!"

The students are asked to hold their answers high in the air while teachers count and record the number of responses for each letter (Figure 15.3). This strategy models the multiple-choice assessment used in traditional unidimensional testing, but uses a kinesthetic community approach to engage the students to recall prior knowledge and think about the best answer. The tallying of student responses permits the visual feedback of the correct answers and permits the visualization of the improvement of the class in a "whole-team" approach. Furthermore, it permits the infusion of mathematics if a teacher wants to introduce graphing skills to track class self-improvement.

Figure 15.3. An example of fifth graders engaged in "P(review)"

Figure 15.4. Fifth-Grade Data Recording Sheet (P)review

Question	Possible Answers	Beginning of lesson (Pre)	End of lesson (Post)	Beginning of next lesson (Post2)
1. Sources of Energy that are endless, meaning we will not run out of them, are called:	A. Renewable			
	B. Semi-renewable			
	C. nonrenewable			
Total Responses				
2. The following is a source of renewable energy:	A. Coal			
	B. Wind			
	C. Oil			
Total Responses				
3. The Industrial Revolution and more burning of fossil fuels to power trains and factories caused green house gases to:	A. Increase			
	B. Decrease			
	C. Stay the same			
Total Responses				

Evidence of Effectiveness

Results in Table 15.1 and Figure 15.5 suggest that students are consistently gaining and retaining state-mandated content knowledge learned through these informal in-class activities. In all lessons students improved their cognitive learning, although physical science concepts in "Light and the Environment" remained the greatest challenge for fifth-grade students.

Table 15.1. Tallying the Fifth-Grade P(review) Data

	Grade 5 lesson 1 – Energy and Climate Change
	% of youth who answered correctly
pre	82%
post	96%
post2	92%
	Grade 5 lesson 2 – Light and Environment
	% of youth who answered correctly
pre	41%
post	67%
post2	74%
	Grade 5 lesson 3 – Water Conservation and Pollution
	% of youth who answered correctly
pre	68%
post	100%
post2	97%
	Grade 5 lesson 4 – West River Memorial Park Exploration
	% of youth who answered correctly
pre	68%
post	80%
post2	

Conversations with teachers suggested that the informal feeling of the kinesthetic musical questioning approach is helping students perform better in multiple-choice questioning compared to traditional multiple-choice practices. In fact, teachers reported that students who did not perform as well on standardized test questions assessing the same knowledge showed the most improvement with the more social HOLA approach. When HOLA was used to review material after the exam, these students performed equally as well, suggesting that while they had retained the information, that fact was not being demonstrated with the traditional testing

Figure 15.5. P(review) Results for the Fifth Grade

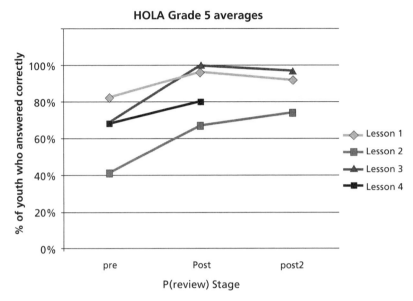

methods. This suggests that such a kinesthetic method needs to be used more by teachers at Barnard, and once confidence has been built, students need help transitioning from this musical/kinesthetic assessment approach to the more standardized approach that will help such students compete equally well on statewide and national levels of testing.

Student quotes about their Barnard experiences support evidence of achievement:

> Barnard is a great school; you learn a lot of things that you are supposed to learn in high school. We go on field trips; it is a really fun school. The most exciting thing we do here is working in the garden planting sunflower seeds in the spring and coming back in the fall to see how large they became (Corey, Grade 6).

> We learn a lot about saving and creating energy. We have gone on a lot of field trips to learn these things (Lorant, grade 8).

> I've been at Barnard for seven years. During that time I've learned about being green and helping our environment. My greatest experience was being part of an article for Dynamath Magazine that had to do with solar energy at Barnard (Michilla, grade 7).

> I've been here for one year, but I've learned a lot about being green. This school is really global and you learn about the community and how to make it a better place (Ralik, grade 8).

Teachers reported that students traveling to the Tilcon Rock Quarry discussed the advantages and disadvantages of mining after reading about the subject, and then discussed these with a quarry foreman. The class was curious about the process of blasting rocks from the ground and then breaking them to the right size to use on roads and buildings. The trip illuminated the finite nature of procuring supplies from the earth (even rocks), and the responsibilities people have to use the resource wisely. Additionally, conversations with "Save the Sound" representatives permitted students to share their thoughts about how fish move upriver. Students gained insights into the purpose of fish ladders, and why often it is necessary for people to help animals in urban areas complete their life cycles. Students gained insights into the purpose of environmental scientists, and the need to help animals when manmade structures, like roads, get in the way. The results in growth of more positive student attitudes are most striking with the gardening program. Students' preconceived attitude about dirt being "yucky" was completely changed to viewing dirt as a nurturing tool with which they can grow their own food. It is common to hear students talk about using rakes and shovels to "till" the soil, "plant" seeds, and "watch" plants grow. They discuss talking to their families about starting gardens at home, and offering ideas to their teachers about what plants to try to grow next. Teaching students to use small (recycled) containers to grow plants helps them visualize how gardening can be done in urban homes and develop subsequent confidence to implement these actions in their community.

Conclusions

A method to measure student understanding of content is to engage them in conversations and listen to how the students process information in response to problem solving of socially relevant issues. This learning strategy retains student engagement inasmuch as the student is actively presenting and thinking about what is being discussed. The student is thinking critically and engaged in logical problem-solving strategies needed in scientific thinking. In order to engage in such activity the students need to be knowledgeable about scientific issues.

In 2001 the No Child Left Behind Act was initiated in response to students' poor school performance in math and reading. Eventually, Connecticut implemented the Science Mastery exams in 2008 to document student performance and growth. Schools must develop techniques that permit all students to learn science in ways that meet the rigor of the curriculum and are meaningful to the student.

The Connecticut State and New Haven District science curricula act as the foundation for the science concepts taught at the school. But, they are approached from varying contexts and student teams. Magnet schools are implemented to attract students to thematic curricula. In these schools, partners are used to help differentiate instruction and encourage students to interact with practitioners and scholars. Instructional methods that also help students master traditional test-taking strategies as well are important vehicles for teaching students to compete globally. The HOLA approach is an example of how kinesthetic assessment can help improve student experiences with multiple-choice test-taking skills while also engaging students with socially important scientific issues.

References

Blumenfeld, P. C., E. Soloway, R. W. Marx, J. S. Krajcik, M. Guzdial, and A. Palincsar. 1991. Motivating project based learning: Sustaining the doing, supporting the learning. *Educational Psychologist* 25: 369-398.

Gardner, H. 1999. *Intelligence reframed: Multiple intelligences for the 21st centu*ry. New York: Basic Books.

Ibarra, R. A. 2001. *Beyond affirmative action: Reframing the context of higher education.* Madison, WI: The University of Wisconsin Press.

King, K. 2007. *Teaching for K–12 scientific literacy. Integrating the national science education standards into classroom practice.* Upper Saddle River, NJ: Pearson Education.

National Research Council (NRC). 1996. *National science education standards.* Washington, DC: National Academies Press.

National Research Council (NRC). 2000. *Educating teachers of science, mathematics and technology: New practices for the new millennium.* Washington, DC: National Academies Press.

Place, G. 2004. Youth recreation leads to adult conservation. *Parks and Recreation* (Feb. 1).

Sanchez-Burks, J., R. E. Nisbett, and O. Ybarra. 2000. Cultural styles, relational schemas, and prejudice against out-groups. *Journal of Personality and Social Psychology* 79 (2): 174–189.

Winking, D., and L. A. Bond. 1995. *Transforming teaching and learning in urban schools through alternative assessment.* Oak Brook, IL: North Central Regional Educational Laboratory.

Endword:

What Is Revealed From Using Content Arising From Personal and Societal Issues While Also Focusing on NSES Goals 2 and 3?

Robert E. Yager
University of Iowa

The 15 chapters in this monograph provide quite different ways to illustrate meeting NSES Goals 2 and 3 while also including interesting ways science content can be approached from personal and social perspectives for PreK to college students. Although there are 38 authors, few are science teachers and typical NSTA members. These authors are impressive and bring many original ideas for the needed reforms in science education. Only three of the chapters are authored by single authors—and even these three make the students major partners in planning and carrying out the programs. In many ways the stories that comprise this Exemplary Science Program monograph (our seventh) are exciting because they do not use typical definitions of content, of goals for teaching school science, of indicators of student learning. Assessments that are used to indicate success are *not* standard achievement measures that seem to be what captures the attention of the public—often causing stress and lack of encouragement to define and use the real parameters of science. Too often the expectation for students is little more than remembering verbatim notes taken during class sessions and information included in textbooks.

Certainly much effort and attention is placed on Goal 1: ensuring that all students experience the richness and excitement of knowing about and understanding the natural world. This is what all practicing scientists do. It also defines what school science should provide as experiences for all students in all classes called science. Such experiences also excite technologists (especially engineers) and encourage them to invent needed materials to make life better (and to serve scientists with better investigative tools). The involvement of engineers and inventors who produce devices that can be used and seen as useful products are too often separated from science itself.

Science and technology both employ questioning, and attempts to find answers and results for the questions; in science we do not know the answers before collecting and analyzing data; but, in technology we always know the results that we strive to attain. Both require thinking and actions. Carl Sagan has stated that all people start out as scientists full of questions, awe, wonderment, and creative ideas (NRC 1998, p.1). There is even evidence that the human brain is at work while in the mother's womb, reacting to heat, pressure, and movement. Preschool children are seemly curious about everything around them. This is not like the negative reactions to science study found in schools. Students typically decrease in their creativity (questioning, evidence collecting, problem resolution, and engaging in corrective actions) the longer they are in school. The most success in "doing" science in schools is at the early elementary level. And yet, typical science courses are devoid of encouraging what all students have before they start school—curiosity. Since new inventions are not listed as something we know, as in the case of science, such activities with technology usually are supplements to the school program or merely fun experiences in summer camps.

The 15 chapters also focus on many projects that supplement the typical school programs and the great dependence on textbooks found there. This added focus helps change the status quo, where it has been reported that in excess of 90% of all science teachers rely on a textbook for 90% of their instruction (Harms and Yager 1981). Even so-called science laboratories are places where students are merely required to follow directions and undertake specific actions. Too often they are little more than a set of directions for illustrating what is in the textbooks.

Another important point is that typical science teaching and the curriculum created to accomplish it require only outlining and imparting what current scientists think they know. Science is rarely described as a "work in progress." As a result, most traditional teaching results in loss of student interest. (And yet, attitudes about science are very positive in early elementary grades—but such positive reactions are nearly gone by grade 12!) They decline even further in colleges, even though it is only the "upper" half of high schoolers who enroll in colleges. Some have argued that typical science teaching should be outlawed if successes of it are to be measured by the degree students "like it." Few find science study interesting and useful! This is not the situation reported in the efforts provided in the 15 chapters of this monograph.

Too many teachers do not use research, logic, and perseverance as ways of changing typical teaching while maintaining dependence on current textbooks. Such textbooks would not exist if books and curricula were related to what is identified in the NSES as the "four" goals for defining what should be accomplished in school science and if it included all parameters identified as part of school science content. The typical school curriculum needs to be framed differently. Most teachers, school administrators, parents, and others equate good teaching as the teacher in control, sharing what he or she knows, and assessing success by what definitions students can recite and from notes of "teacher talk" or from reading the textbook. All of this is done too often by repeating only what is in textbooks and teacher lesson plans.

Typical science programs and teaching are good for some activities, but most are alien to real science. Students in science should be having fun, with their minds engaged and becoming increasingly more curious. There should be ways that community leaders can assist with the work students do in their communities. Such endeavors should be indicators of learning while providing real experiences with actually "doing" science. This is basic to the reforms envisioned in the NSES.

Much has been written about reforms. But, writing does little to produce change. Many call for defining science as agreements concerning treatments of the "Big Ideas." Many call for specific "required" assessments nationally. We hope that the Exemplary Science Program efforts provide examples of people (teachers, students, and others) having fun and being enthused with experiencing real science. Both teachers and students can become more anxious to stimulate real learning, community betterment(s), and more participation in society generally and certainly beyond the school walls. More creativity and more positive attitudes are generally considered fine outcomes; they are just not seen as a clear focus in many science classrooms. In fact research reveals that the outcomes regarding attitudes and creativity actually worsen the longer students in K–12 experience typical school science. Few find fault with engaged students, who are able to use their learning, and who help resolve societal and personal issues. But, too few students are encouraged (or even enabled) to be so motivated. This is the excitement of what is reported in the chapters of this monograph.

In all cases, student-centeredness is a desirable situation in science classrooms. But how do we really achieve it? Teachers must do less with lesson planning and developing schedules for students to follow rigidly. Students generally like the diversity that can be encouraged as they see the results of their actions and their learning. Perrone (1994) has offered a fine list of ways that intellectual engagement can be achieved in science classes. These are ways that illustrate students as cornerstones in achieving real learning. In his eight examples,

1. students help define the content;

2. students have time to wonder and to find a particular direction that interests them;

3. topics have a "strange" quality—something common seen in a new way, evoking a lingering question;

4. teachers permit—even encourage—different forms of expression and respect students' views;

5. teachers are passionate about their work; the richest activities are those "invented" by the teachers and their students;

6. students create original and public products; they gain some form of *expertness*;

7. students *do* something (such as participate in political actions, write letters to the editor, or work with the homeless); and

8. students sense that the results of their work are not predetermined or fully predictable.

Many of these conditions are exemplified by the program described in this monograph.

We also hope that student-centeredness means removing teachers from the front of the room, talking and directing. What better can happen than to have all teachers become learners *with* their own students? Students in all classrooms have always had the power to make the most basic choices about their learning: They may choose to engage in learning or to disengage. Teachers in tune with the NSES reforms aim to inspire students to choose "engagement." In short, we want students to choose to learn in order to learn. Too often students are merely expected to do as they are told and threatened with poorer grades if they do not conform. Starnes and Paris (2000) have worked extensively on the importance (perhaps necessity) that teachers help (not require) their students to learn instead of illustrating their use and skills by merely defining science constructs teachers want students to learn. Students must be involved and excited with the chance to provide evidence of their learning. The primary mode of assessment in science classrooms should not be determined by assessment experts with teachers merely requiring it because it is expected by administrators, parents, and government leaders.

The diverse group of authors for the 15 chapters also indicates collaborative efforts and involvement of teachers as a part of the projects and instructional teams. Noah Feinstein (2009) has recently indicated a problem with science programs not being clear on the goals and how to realize them. He identifies these two issues as follows:

> American science education is plagued by a fundamental confusion—a mismatch between goals we claim to value and the strategies we use to achieve them. This confusion is rooted in the seemingly simple idea that science education should prepare students for the future.

> For some, this "preparation" is about good citizenship and a satisfying life. For others, it's about a scientifically and technically skilled workforce. Each vision is clear and compelling, but each requires a different educational strategy. To prepare citizens, we should base our educational strategy on an understanding of everyday life—in particular, the role played by science in everyday life. To prepare a scientific workforce, we should ensure that students have the skills and knowledge to pursue advanced training in scientific and technical fields. We claim to value civic engagement and science literacy, but our education system is more suited to producing a scientific workforce. Because we like the idea that science education prepares students for the future, we take for granted that any good science education can make us capable of leading "interesting, responsible, and productive lives." Yet as hard as we try, we will never reach citizenship goals with a workforce teaching strategy.

Not only does Feinstein provide an interesting look at our 15 chapters, he also introduces our plans for the next Exemplary Science Program monograph: a focus on science and technology careers. Some of the concerns and correctives are illustrated in the 15 chapters of this monograph. Some of them suggest a deeper focus for our new plan. This discussion point is the importance of achieving science literacy for all and what it really means. Morris Shamos (1995) has written a book called *The Myth of Scientific Literacy*. It is important to note that Shamos, a physicist, confessed that he was "scientifically" illiterate. He was quick to admit that he could not pick up each issue of *Science* (the journal of the American Association for the Advanced of Science) and read and understand every article. And yet, too often we expect all K–12 students to know everything in a yearlong course. Some have even argued that textbooks should not be allowed in school science—since they constrict science to merely taking what is given in the name of science without students ever experiencing or learning real science.

The important thing for readers of this monograph is to focus on the reported results and the evidence of success for student learning reported in each chapter. Evidence was to include at least a third of the content for each chapter. Are new goals articulated? Did our primary focus on Goals 2 and 3 become a distraction? Was our focus only on one of the new facets of science content successful in drawing attention to possible use of personal and societal issues as organizers and motivators? You as a reader are invited to judge. What ideas started your thinking? Can we carry on further dialogue about the exciting programs included? They need not be an ending— but hopefully a beginning of more discussions, more collaboration, and more real successes with reforms. Perhaps Rutherford and Ahlgren (1990) were right in recognizing that real educational reform takes much time. It was their rationale with Project 2061, the date that Halley's Comet will be seen again from Earth. It is certainly true that the needed changes will not come from attending a workshop or two. Change, though needed, cannot be realized easily or quickly. But, must it take a lifetime to succeed with such fundamental changes? The Exemplary Science Program team is involved and supportive because there is progress and hope that the process can be speeded and that more successes will come by sharing the efforts of the authors of these monograph chapters. As this monograph was going to press, we were diligently interacting with others who are committed to enticing more students to become immersed in efforts involving both science and technology in terms of careers and better living as contributing citizens.

References

American Association for the Advancement of Science (AAAS). 1989. *Project 2061: Science for all Americans*. New York: Oxford University Press.

American Association for the Advancement of Science (AAAS). 1990. *Science for all Americans*. New York: Oxford University Press.

Feinstein, N. 2009. Prepared for what? Why teaching "everyday science" makes sense. *Phi Delta Kappan* 90 (10): 762–766.

Harms, N. C., and R. E. Yager, eds. 1981. *What research says to the science teacher, vol. 3*. Washington, DC: NSTA Press.

National Research Council (NRC). 1998. *Every child a scientist: Achieving scientific literacy for all*. Washington, DC: National Academies Press.

Perrone, V. 1994. How to engage students in learning. *Educational Leadership* 51 (5): 11–13.

Rutherford, F. J., and A. Ahlgren. 1990. *Science for all Americans? A Project 2061 report on literacy goals in science, mathematics, and technology*. New York: Oxford University Press.

Shamos, M. H. 1995. *The myth of scientific literacy*. Piscataway, NJ: Rutgers University Press.

Starnes, B. A., and C. Paris. 2000. Choosing to learn. *Phi Delta Kappan* 81 (1): 392–397.

Contributors

Hakan Akcay, author of *Applications of Biology as Part of a Pre-Service Program for Science Teachers,* is an instructor of science education at Marmara University, Istanbul, Turkey.

Scott Applebaum, coauthor of *Using Socioscientific Issues as Context for Teaching Concepts and Content,* is a science instructor of biology, anatomy, and physiology at Palm Harbor University High School in Palm Harbor, Florida.

Anthony Bartley, coauthor of *Issues-Based Learning and Inquiry in Environmental Science: Meeting the Third Goal of the National Science Education Standards,* is a faculty of education member at Lakehead University, Thunder Bay, Ontario, Canada.

Oksana Bartosh, coauthor of *Tahoma Outdoor Academy: Learning about Science and the Environment Inside and Outside the Classroom,* is a researcher at the Canadian Council on Learning in Vancouver, BC, Canada.

Erin Baumgartner, coauthor of *Your Students as Scientists: Guidelines for Teaching Science through Disciplinary Inquiry,* is an assistant professor of biology at Western Oregon University, Monmouth, Oregon.

David L. Brock, author of *Securing a "Voice:" The Environmental Science Summer Research Experience for Young Women,* is the E.S.S.R.E. project director at Roland Park Country School, Baltimore, Maryland.

Clara Cahill, coauthor of *Engaging Students in Content Learning and Scientific Critique Through a Nanscience Context,* is a graduate student at the University of Michigan, Ann Arbor, Michigan.

Kristy Loman Chiodo, coauthor of *Using Socioscientific Issues as Context for Teaching Concepts and Content,* is a marine biology teacher at Robinson High School, Tampa, Florida.

Kabba E. Colley, coauthor of *Project-Based After-School Science in New York City,* is chief evaluation and research officer of Eduinformatics in Plainfield, Vermont.

Cesar Delgado, coauthor of *Engaging Students in Content Learning and Scientific Critique Through a Nanoscience Context,* is an assistant professor at The University of Texas at Austin, Austin, Texas.

Gail Dickinson, coauthor of *Developing Expertise in Project-Based Science: A Longitudinal Study of Teacher Development and Student Perceptions,* is an assistant professor at the Texas State University-San Marcos, Texas.

Marjorie Drucker, coauthor of *An "HOLA" Approach to Learning Science,* is the science theme coordinator at Barnard Environment Studies Magnet School, New Haven Public Schools, New Haven, Connecticut.

Kathleen A. Fadigan, coauthor of *The CHANCE Program: Transitioning From Simple Inquiry-Based Learning to Professional Science Practice,* is an assistant professor of science education at The Pennsylvania State University, Abington, Pennsylvania.

David Fortus, coauthor of *Developing Students' Sense of Purpose with a Driving Question Board,* is a senior scientist at the Weizmann Institute of Science in Rehovot, Israel.

Barbara Hug, coauthor of *Linking Science, Technology, and Society by Examining the Impact of Nanotechnology on a Local Community,* is a clinical assistant professor at the University of Illinois, Champaign, Illinois.

Julie K. Jackson, coauthor of *Developing Expertise in Project-Based Science: A Longitudinal Study of Teacher Development and Student Perceptions,* is an assistant professor at Texas State University-San Marcos, Texas.

Doug Jones, coauthor of *Issues-Based Learning and Inquiry in Environmental Science: Meeting the Third Goal of the National Science Education Standards,* is head of the science department at Sir Winston Churchill C&VI High School, Thunder Bay, Ontario, Canada.

Juan D. López, Jr., coauthor of *"Who Ate Our Corn?" We Want To Know And So Should You!,* is a research entomologist at the USDA-ARS, Areawide Pest Management Research Unit in College Station, Texas.

Jolie Mayer-Smith, coauthor of *Tahoma Outdoor Academy: Learning about Science and the Environment Inside and Outside the Classroom,* is an associate professor at the University of British Columbia, Vancouver, Canada.

Jacqueline S. McLaughlin, coauthor of *The CHANCE Program: Transitioning From Simple Inquiry-Based Learning to Professional Science Practice,* is an associate professor of biology and CHANCE Founding Director at Penn State Lehigh Valley, Center Valley, Pennsylvania.

Wayne Melville, coauthor of *Issues-Based Learning and Inquiry in Environmental Science: Meeting the Third Goal of the National Science Education Standards,* is an associate professor in science education at Lakehead University, Thunder Bay, Ontario, Canada.

Joseph Muskin, coauthor of *Linking Science, Technology, and Society by Examining the Impact of Nanotechnology on a Local Community,* is an educator coordinator at the University of Illinois, Urbana, Illinois and a science teacher at the Next Generation School, Champaign, Illinois.

Andrew J. Petto, author of *Communic—Able: Writing to Learn About Emerging Diseases,* is senior lecturer in anatomy and physiology at the University of Wisconsin-Milwaukee, Milwaukee, Wisconsin.

Jason Pilot, coauthor of *Issues-Based Learning and Inquiry in Environmental Science: Meeting the Third goal of the National Science Education Standards,* is a high school science teacher at Sir Winston Churchill Collegiate and Vocational Institute, Thunder Bay, Ontario, Canada.

Theodora Pinou, coauthor of *An "HOLA" Approach to Learning Science,* is an associate professor at Western Connecticut State University, Danbury, Connecticut.

Wesley B. Pitts, coauthor of *Project-Based After-School Science in New York City,* is an assistant **professor** of science education in the department of middle and high school education at Lehman College, City University of New York in Bronx, New York.

Amy E. Ryken, coauthor of *Tahoma Outdoor Academy: Learning about Science and the Environment Inside and Outside the Classroom,* is an associate professor at the University of Puget Sound in Tacoma, Washington.

Timothy P. Scott, coauthor of *"Who Ate Our Corn?" We Want To Know And So Should You!,* is associate dean at College of Science—Texas A&M University, College Station, Texas.

Kanesa Duncan Seraphin, coauthor of *Students as Scientists: Guidelines for Teaching Science through Disciplinary Inquiry,* is an assistant professor at the University of Hawaii's Curriculum Research & Development Group (CRDG) & director of the Hawaii Sea Grant Center for Marine Science Education.

Yael Shwartz, coauthor of *Developing Students' Sense of Purpose With a Driving Question Board,* is a senior scientist at the Weizmann Institute of Science in Rehovot, Israel.

Minyoung Song, coauthor of *Engaging Students in Content Learning and Scientific Critique Through a Nanoscience Context,* is a PhD student at the University of Michigan, Ann Arbor, Michigan.

Elizabeth E. Studley, coauthor of *An "HOLA" Approach to Learning Science,* is the program director at Solar Youth, Inc., New Haven, Connecticut.

Emily J. Summers, coauthor of *Developing Expertise in Project-Based Science: A Longitudinal Study of Teacher Development and Student Perceptions,* is an assistant professor at Texas State University-San Marcos, Texas.

Margaret Tudor, coauthor of *Tahoma Outdoor Academy: Learning About Science and the Environment Inside and Outside the Classroom,* is executive director of the Pacific Education Institute, Olympia, Washington.

Janet Wattnem, coauthor of *Linking Science, Technology, and Society by Examining the Impact of Nanotechnology on a Local Community,* is high school teacher at Mahomet-Seymour High School, Mahomet, Illinois.

Ayelet Weizman, coauthor of *Developing Students' Sense of Purpose With a Driving Question Board,* is a senior researcher at the Weizmann Institute of Science in Rehovot, Israel.

Craig Wilson, coauthor of *"Who Ate Our Corn?" We Want To Know And So Should You!,* is a senior research associate at College of Science—Texas A&M University, College Station, Texas.

Dana L. Zeidler, coauthor of *Using Socioscientific Issues as Context for Teaching Concepts and Content,* is a secondary science education professor at the University of South Florida, Tampa, Florida.

Index